Algebra: A Comprehensive Course

Algebra: A Comprehensive Course

Kevin Houston

NYRESEARCH
PRESS

New York

Published by NY Research Press
118-35 Queens Blvd., Suite 400,
Forest Hills, NY 11375, USA
www.nyresearchpress.com

Algebra: A Comprehensive Course
Kevin Houston

International Standard Book Number: 978-1-63238-876-6 (Hardback)

Cataloging-in-Publication Data

Algebra : a comprehensive course / Kevin Houston.
    p. cm.
Includes bibliographical references and index.
ISBN 978-1-63238-876-6
1. Algebra. 2. Mathematical analysis. 3. Mathematics. I. Houston, Kevin.
QA155 .A44 2022
512--dc23

TABLE OF CONTENTS

This book is a culmination of my many years of practice in this field. I attribute the success of this book to my support group. I would like to thank my parents who have showered me with unconditional love and support and my peers and professors for their constant guidance.

Algebra is primarily concerned with the study of mathematical symbols as well as the rules that operate such symbols. It is applied in most of the sub-domains within mathematics. Algebra makes use of letters to denote numerical values. Some of the major branches of algebra are elementary algebra and abstract algebra. Elementary algebra focuses on the study of variables and polynomials. Abstract algebra studies the abstraction such as groups, rings and fields as well as elementary equation solving. It is applied in the study of various fields such as algebraic topology and algebraic number theory. This book is compiled in such a manner, that it will provide in-depth knowledge about the theory and practice of algebra. Some of the diverse topics covered herein address the varied branches that fall under this category. Coherent flow of topics, student-friendly language and extensive use of examples make this book an invaluable source of knowledge.

The details of chapters are provided below for a progressive learning:

Chapter – What is Algebra?

Algebra is the branch of mathematics which focuses on the study of mathematical symbols along with the rules for the manipulation of these symbols. This is an introductory chapter which will introduce briefly all the different operations in algebra such as addition, multiplication, subtraction and division.

Chapter – Branches of Algebra

Algebra is divided into different branches namely abstract algebra, elementary algebra, linear algebra, universal algebra, Boolean algebra, etc. This chapter closely examines these different branches of algebra to provide an extensive understanding of the subject.

Chapter – Algebraic Expressions

Algebraic expressions consist of variables, constants and other algebraic operations such as addition, subtraction, etc. There are three main types of algebraic expressions which are monomial expression, binomial expression and polynomial expression. This chapter has been carefully written to provide an easy understanding of these algebraic expressions.

Chapter – Algebraic Functions and Equations

Algebraic functions can be classified into linear function, quadratic function, cubic function, quartic function, etc. Algebraic equations include linear equation, quadratic equation, cubic equation, quantic equation, etc. All these algebraic functions and equations have been carefully analyzed in this chapter.

Chapter – Algebraic Number Theory

Algebraic number theory studies the integers, rational numbers and their generalizations with the use of abstract algebra. Class field theory and abstract analytic number theory are some of the theories that fall under its domain. This chapter discusses in detail these theories related to algebra.

Chapter – Algebraic in Theorems

There are a number of important theorems used in algebra such as Abel–Ruffini theorem, Amitsur-Levitzki theorem, Bernstein-Kushnirenko theorem, Hilbert's basis theorem, remainder theorem and factor theorem. The topics elaborated in this chapter will help in gaining a better perspective about these theorems in algebra.

Kevin Houston

CHAPTER 1

What is Algebra?

Algebra is the branch of mathematics which focuses on the study of mathematical symbols along with the rules for the manipulation of these symbols. This is an introductory chapter which will introduce briefly all the different operations in algebra such as addition, multiplication, subtraction and division.

Algebra is a branch of mathematics that substitutes letters for numbers, and an algebraic equation represents a scale where what is done on one side of the scale is also done to the other side of the scale and the numbers act as constants. Algebra can include real numbers, complex numbers, matrices, vectors, and many more forms of mathematic representation.

The field of algebra can be further broken into basic concepts known as elementary algebra or the more abstract study of numbers and equations known as abstract algebra, where the former is used in most mathematics, science, economics, medicine, and engineering while the latter is mostly used only in advanced mathematics.

Properties of Algebra

The properties involved in algebra are as follows:

Commutative Property of Addition

Changing the order of addends does not change the sum. The addends may be numbers or expressions. That is $(a + b) = (b + a)$ where a and b are any scalar.

Example:

Consider the real numbers 5 and 2.

Obtain the value of Left Hand Side (LHS) of the rule.

$$a + b = 5 + 2$$
$$= 7$$

Obtain the value of Right Hand Side (RHS) of the rule.

$$b + a = 2 + 5$$
$$= 7$$

Since the sum is same, the commutative property holds for addition.

Example:

Consider the algebraic expression $x^2 + 2x$ where $x \in \mathbb{R}$.

$$x^2 + x = x + x^2$$

Substitute $x = -1$ on both sides.

$$(-1)^2 + (-1) = (-1)(-1)^2$$
$$1 + (-1) = (-1) + 1$$
$$0 = 0$$

Since the sum is same, the commutative property holds for addition.

Commutative Property of Multiplication

Changing the order of factors does not change the product. The factors may be numbers or expressions. That is, $(a \times b) = (b \times a)$.

Example:

Consider the real numbers 15, −2.

Obtain the value of Left Hand Side (LHS) of the rule.

$$a \times b = 15 \times (-2)$$
$$= -30$$

Obtain the value of Right Hand Side (RHS) of the rule.

$$b \times a = (-2) \times 15$$
$$= -30$$

Since the product is same, the commutative property holds for multiplication.

Example:

Consider the algebraic expression $(x^2 + 2x)(2x + 1)$ where $x \in \mathbb{R}$.

$$(x^2 + 2x)(2x + 1) = (2x + 1)(x^2 + 2x)$$

substitute $x = 2$ on both side

$$(2^2 + 2(2))(2(2) + (1)) = (2(2) + 1)(2^2 + 2(2))$$
$$(4 + 4)(4 + 1) = (4 + 1)(4 + 4)$$
$$(8)(5) = (5)(8)$$
$$40 = 40$$

Since the product is same, the commutative property holds for multiplication.

Associativity Property of Addition and Multiplication

The associate property defines that grouping of more than two numbers and performing the basic arithmetic operations of addition and multiplication does not affect the final result. Note that grouping means placing the parenthesis.

Rule for Addition

If a, b, and c are any numbers, then $(a + b) + c = a + (b + c)$ holds true.

Example:

Consider $a = -2$, $b = 4$, and $c = 5$.

Obtain the value of Left Hand Side (LHS) of the rule.

$$\begin{aligned}(a+b)+c &= (-2+4)+5 \\ &= 2+5 \\ &= 7\end{aligned}$$

Obtain the value of Right Hand Side (RHS) of the rule.

$$\begin{aligned}a+(b+c) &= -2+(4+5) \\ &= -2+9 \\ &= 7\end{aligned}$$

Since the sum is same, the associative property of addition holds true. Therefore, it can be concluded that the grouping of numbers in any order does not change the sum.

Example:

Consider the algebraic expression, $x, y, 2x$ where $x, y\, x \in \mathbb{R}$.

$(x+y)+2x = x+(y+2x)$ is holds for addition.

Rule for Multiplication

If a, b, and c are any numbers, then $(a\times b)\times c = a\times(b\times c)$ holds true.

Example:

Consider $a = -2$, $b = 4$, and $c = 5$.

Obtain the value of Left Hand Side (LHS) of the rule.

$$\begin{aligned}(a\times b)\times c &= ((-2)\times 4)(\times 5) \\ &= (-8)\times 5 \\ &= -40\end{aligned}$$

Obtain the value of Right Hand Side (RHS) of the rule.

$$a \times (b \times c) = (-2) \times (4 \times 5)$$
$$= (-2) \times 20$$
$$= -40$$

Example:

Consider the algebraic expression $(x+2), (x^2), (2x)$ where $x \in \mathbb{R}$.

$((x+2) \times x^2) \times 2x = (x+2) \times (x^2 \times 2x)$ is holds for multiplication.

Distributive Property

The distributive property defines that the product of a single term and a sum or difference of two or more terms inside the bracket is same as multiplying each addend by the single term and then adding or subtracting the products.

1. If $a \cdot (b+c) = (a \cdot b) + (a \cdot c)$, then it is known as left distributive law.

2. If $(a+b) \cdot c = (a \cdot c) + (b \cdot c)$, then it is known as right distributive law.

More generally, the property is true for number of addends.

$$a(b+c+d+\cdots) = ab + ac + ad + \cdots$$
$$a(b-c-d-\cdots) = ab - ac - ad - \cdots$$

Rule for Multiplication Over Addition

If a, b and c are any numbers, then $a \cdot (b+c) = a \cdot b + a \cdot c$ holds true. The distribution can be done in two ways, namely left distribution and right distribution. That is,

1. If $a.(b+c) = (a \cdot b) + (a \cdot c)$, then it is known as left distributive law.

2. If $(a+b) \cdot c = (a \cdot c) + (b \cdot c)$, then it is known as right distributive law.

Example:

Consider a = 2, b = 4, and c = 9.

Obtain the value of Left Hand Side(LHS) of the rule.

$$a \cdot (b+c) = 2 \cdot (4+9)$$
$$= 2(13)$$
$$= 26$$

Obtain the value of Right Hand Side (RHS) of the rule.

$$a \cdot b + a \cdot c = 2 \cdot 4 + 2 \cdot 9$$
$$= 8 + 18$$
$$= 26$$

Since the result is same, the distributive property for multiplication over addition holds true.

That is, $2 \cdot (4 + 9) = 2 \cdot 4 + 2 \cdot 9$.

Rule for Multiplication Over Subtraction

If a, b and c are any numbers, then $a \cdot (b - c) = a \cdot b - a \cdot c$ holds true. Similarly over to the addition rule, the distribution for multiplication over subtraction can be done in two ways, namely left distribution and right distribution. That is,

1. If $a \cdot (b - c) = (a \cdot b) - (a \cdot c)$, then it is known as left distributive law.
2. If $(a - b) \cdot c = (a \cdot c) - (b \cdot c)$, then it is known as right distributive law.

Example:

Consider a = 2, b = 4, and c = 9.

Obtain the value of Left Hand Side (LHS) of the rule.

$$a \cdot (b - c) = 2 \cdot (4 - 9)$$
$$= 2(-5)$$
$$= -10$$

Obtain the value of Right Hand Side (RHS) of the rule.

$$a \cdot b - a \cdot c = (2 \cdot 4) - (2 \cdot 9)$$
$$= 8 - 18$$
$$= -10$$

Since the result is same, the distributive property for multiplication over subtraction holds true. That is, $2 \cdot (4 - 9) = 2 \cdot 4 - 2 \cdot 9$.

Example:

Consider the algebraic expression $3x, x^2, x^3 \in \mathbb{R}$.
$$3x(x^2 + x^3) = 3x(x^2) + 3x(x^3) = 3x^3 + 3x^4$$
$$3x(x^2 - x^3) = 3x(x^2) - 3x(x^3) = 3x^3 - 3x^4$$

The property holds for both addition and subtraction.

Additive Identity Property

The additive identity is zero. That is, the sum of any number and zero is the same number. $a + 0 = 0 + a$.

Example:

Let 5 be a real number such that 5 + 0 = 5.

Let $x \in \mathbb{R}$ such that $x + 0 = x$.

The Multiplicative Identity Property

The multiplicative identity is 1. That is, the product of any number and 1 is same number, $a \times 1 = 1 \times a = a$.

Example:

Let 6 be a real number such that $6 \times 1 = 6$

Let $x \in \mathbb{R}$ such that $x \times 1 = x$.

Additive Inverse Property

The additive inverse of a is $-a$.

For every real number, $a + (-a) = 0 = (-a) + a$.

Example:

The additive inverse of -7 is $-(-7) = 7$. That is, $-7 + (7) = 0$.

Multiplicative Inverse Property

The multiplicative inverse of a non-zero real number a is $\dfrac{1}{a}$. That is, $a \cdot \left(\dfrac{1}{a} \right) = 1$

Example:

The multiplicative inverse of 2 is $\dfrac{1}{2}$.

A combination of variables, constants, and operators constitute an algebraic expression. The four basic operations of mathematics viz. addition, subtraction, multiplication, and division can also be performed on algebraic equations or expressions. Addition and subtraction of algebraic expressions are almost similar to addition and subtraction of numbers. But in the case of algebraic expressions, like terms and the unlike terms must be sorted together.

The terms whose variables and their exponents are same are known as like terms and the terms having different variables are unlike terms.

The terms whose variables and their exponents are same are known as like terms and the terms having different variables are unlike terms.

Example: $-5x^2 + 12xy - 3y + 7x^2 + xy$

In the given algebraic expression, $-5x^2$ and $7x^2$ are like since both the terms have x^2 as the common variable. Similarly, $12xy$ and xy are like terms.

The knowledge of like and unlike terms is crucial while studying addition and subtraction of algebraic expressions because the operation of addition and subtraction can only be performed on like terms.

Addition and subtraction of Algebraic Expressions:

- Addition of Algebraic expressions

For adding two or more algebraic expression the like terms of both the expressions are grouped together. The coefficients of like terms are added together using simple addition techniques and the variable which is common is retained as it is. The unlike terms are retained as it is and the result obtained is the addition of two or more algebraic expressions.

Example: Add $5xy - 3x^2 - 12y + 5x$, $xy - 3x - 12yz + 5x^3$ and $y - 6x^2 - zy + 5x^3$

Solution: For adding these three algebraic expressions the like terms are grouped together and added as shown below.

$$
\begin{array}{rrrrrr}
+5\,xy & -3\,x^2 & -12\,y & +5\,x & +0\,yz & +0\,x^3 \\
+\,xy & +0\,x^2 & +0\,y & -3\,x & -12\,yz & +5\,x^3 \\
+0\,xy & -6\,x^2 & +1\,y & +0\,x & -\,yz & +5\,x^3 \\
\hline
6\,xy & -9\,x^2 & -11\,y & +2\,x & -13\,yz & +10\,x^3
\end{array}
$$

Arrange the terms of the given expressions in the same order.

- Subtraction of algebraic expressions

For subtracting two or more algebraic expressions, it's a better practice to write the expressions to be subtracted below the expression from which it is to be subtracted from. Like terms are placed below each other. The sign of each term which is to be subtracted is reversed and then the resulting expression is added normally.

Example: Subtract $x^2y - 2x^2 - zy + 5$ and $-3x^2 + 3x^3$ from $y^3 + 3x^2y - 6x^2 - 6zy + 7x^3$

Solution: The like terms of the expressions $x^2y - 2x^2 - zy + 5$ and $- 3x^2 + 3x^3$ are written below the like terms of the expression $y^3 + 3x^2y - 6x^2 - 6zy + 7x^3$.

$$
\begin{array}{rrrrrrr}
y^3 & +3x^2y & -6x^2 & -6zy & +7x^3 & +0 \\
-(0y^3 & +x^2y & -2x^2 & -zy & +0x^3 & +5) \\
-(0y^3 & +0x^2y & -3x^2 & +0zy & +3x^3 & +0) \\
\hline
y^3 & +2x^2y & -x^2 & -5zy & +4x^3 & -5
\end{array}
$$

Multiplication is simply repeated addition. We multiply variables and constants in an algebraic expression. For example, the area of a rectangular room is the product of length and breadth. The value of area depends on the value chosen for length and breadth. Similarly volume is the product of length, breadth and height.

Multiplication of Monomial by Monomial

1. Multiply $5x$ with $21y$ and $32z$.

Solution: $5x \times 21y \times 32z = 105xy \times 32z = 3360\,xyz$.

We multiply the first two monomials and then the resulting monomial to the third monomial.

2. Find the volume of a cuboid whose length is 5ax, breadth is 3by and height is 10cz.

Solution: Volume = length × breadth × height

Therefore, volume = $5ax \times 3by \times 10cz =$

$$5 \times 3 \times 10 \times (ax) \times (by) \times (cz) = 150axbycz.$$

Multiplying a Monomials and Polynomials:

$$4ax(2a^2 + 9a + 10) = (4a \times 2a^2) + (4a \times 9a) + (4a \times 10) = 8a^3 + 36a^2 + 40a$$

3. Simplify the below algebraic expression and obtain its value for x = 3.

$$x(x-2) + 5$$

Solution: $x(x-2) + 5,\ x = 3$.

Substituting the value of $x = 3$.

$$3 \times (3-2) + 5 = 3(1) + 5 = 8.$$

4. Simplify the below algebraic expression and obtain its value for y = −1.

$$4y(2y-6) - 3(y-2) + 20$$

Solution: $4y(2y-6) - 3(y-2) + 20$ for $y = -1$.

Substituting the value of $y = -1$.

$$4 \times -1((2 \times -1) - 6) - 3(-1-2) + 20$$
$$= -4(-2-6) - 3(-3) + 20$$
$$= 32 + 9 + 20 = 61$$

Branches of Algebra

Algebra is divided into different branches namely abstract algebra, elementary algebra, linear algebra, universal algebra, Boolean algebra, etc. This chapter closely examines these different branches of algebra to provide an extensive understanding of the subject.

ABSTRACT ALGEBRA

Abstract algebra is a broad field of mathematics, concerned with algebraic structures such as groups, rings, vector spaces, and algebras.

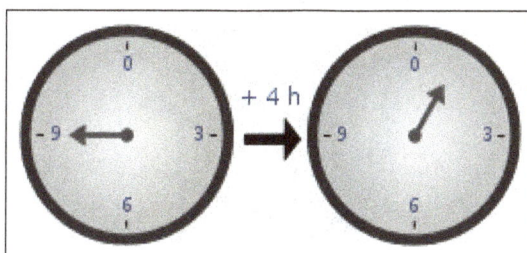

On the 12-hour clock, 9+4=19+4=1, rather than 13 as in usual arithmetic.

Roughly speaking, abstract algebra is the study of what happens when certain properties of number systems are abstracted out; for instance, altering the definitions of the basic arithmetic operations result in a structure known as a ring, so long as the operations are consistent.

For example, the 12-hour clock is an example of such an object, where the arithmetic operations are redefined to use modular arithmetic (with modulus 12). An even further level of abstraction--where only one operation is considered--allows the clock to be understood as a group. In either case, the abstraction is useful because many properties can be understood without needing to consider the specific structure at hand, which is especially important when considering the relationship(s) between structures; the concept of a group isomorphism is an example.

Levels of Abstraction in Abstract Algebra

It is possible to abstract away practically all of the properties found in the "usual" number systems, the tradeoff being that the resulting object--known as a magma (which consists of a set and a binary operation, that need not satisfy any properties other than closure)--is simply too general to be interesting. On the other extreme, it is possible to abstract out practically no properties, which allows for many results to be found, but the resulting object (the usual number systems) is too specific to solve more general problems.

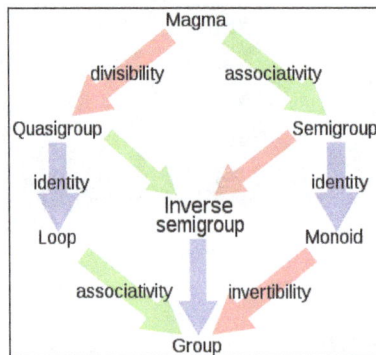

Most of abstract algebra is dedicated to objects that have a reasonable balance between generality and structure, most notably groups and rings in which most of the basic properties of arithmetic are maintained, but their specifics are left free. Still, some higher levels of abstraction are occasionally useful; quasigroups, for instance, are related to Latin squares, and monoids are often used in computer science and are simple examples of categories.

Group Theory

The possible moves on a Rubik's cube form a (very large) group.

Group theory is useful as an abstract notion of symmetry, which makes it applicable to a wide range of areas: the relationship between the roots of a polynomial (as in Galois theory) and the solution methods to the Rubik's cube are both prominent examples.

Informally, a group is a set equipped with a binary operation ∘, so that operating on any two elements of the group also produces an element of the group. For example, the integers form a group under addition, and the nonzero real numbers form a group under multiplication.

The ∘ operation needs to satisfy a number of properties analogous to the ones it satisfies for these "normal" number systems: it should be associative (which essentially means that the order of operations doesn't matter), and there should be an identity element (0 in the first example above, and 1 in the second). More formally, a group is a set equipped with an operation · such that the following axioms hold; note that · does not necessarily refer to multiplication; rather, it should be viewed as a function on two variables (indeed, · can even refer to addition):

Group Axioms

1. Associativity. For any $x, y, z \in G$, we have $(x \cdot y) \cdot z = x \cdot (y \cdot z)$.

2. Identity. There exists an $e \in G$, such that $e \cdot x = x \cdot e = x$ for any $x \in G$. We say that e *is an identity element of G.*

3. Inverse. For any $x \in G$, there exists a $y \in G$ such that $x \cdot y = e = y \cdot x$. We say that y *is an inverse of x.*

It is also worth noting the closure axiom for emphasis, as it is important to verify closure when working with *subgroups (groups contained entirely within another):*

4. Closure: For any $x, y \in G, x * y$ is also in G.

Additional examples of groups include:

- $\mathbb{Z}_n$, the set of integers $\{0, 1, \ldots, n-1\}$ with the operation addition modulo n.

- S_n, the set of permutations of $\{1, 2, \ldots, n\}$ with the operation of composition.

S_3 is worth special note as an example of a group that is not commutative, meaning that a $a \cdot b = b \cdot a$ does not generally hold. Formally speaking, S_3 is nonabelian (an abelian group is one in which the operation is commutative). When the operation is not clear from context, groups are written in the form (set, op); e.g. the nonzero reals equipped with multiplication can be written as ($\mathbb{R}$ *, ·).

Much of group theory (and abstract algebra in general) is centered around the concept of a group homomorphism, which essentially means a mapping from one group to another that preserves the structure of the group. In other words, the mapping of the product of two elements should be the same as the product of the two mappings; intuitively speaking, the product of two elements should not change under the mapping. Formally, a homomorphism is a function $\phi : G \rightarrow H$ such that

$$\phi(g_1) \cdot {}_H \phi(g_2) = \phi(g_1 \cdot {}_G g_2),$$

Where $\cdot_H$ is the operation on H and $\cdot_G$ is the operation on G. For example, $\phi(g) = g \pmod{n}$ is an example of a group homomorphism from $\mathbb{Z}$ to $\mathbb{Z}_n$. The concept of potentially differing operations is necessary; for example, $\phi(g) = e^g$ is an example of a group homomorphism from $(\mathbb{R},) +$ to $(\mathbb{R}^*, \cdot)$.

Ring Theory

Rings are one of the lowest level of abstraction, essentially obtained by overwriting the addition and multiplication functions simultaneously (compared to groups, which uses only one operation). Thus a ring is--in some sense--a combination of multiple groups, as a ring can be viewed as a group over either one of its operations. This means that the analysis of groups is also applicable to rings, but rings have additional properties to work with (the tradeoff being that rings are less general and require more conditions).

The definition of a ring is similar to that of a group, with the extra condition that the distributive law holds as well:

A ring is a set R together with two operations $+$ and $\cdot$ satisfying the following properties (ring axioms):

1. R is an abelian group under addition. That is, R is closed under addition, there is an additive identity (called 0), every element $a \in R$ has an additive inverse $-a \in R$, and addition is associative and commutative.

2. R is closed under multiplication, and multiplication is associative: $\forall a, b \in R$ $a.b \in R$ $\forall a$, $b, c \in R$ $a \cdot (b \cdot c) = (a \cdot b) \cdot c$.

3. Multiplication distributes over addition: $\forall a, b, c \in R a \cdot (b + c) = a \cdot b + a \cdot c$ and $(b + c) \cdot a =$ $= b \cdot a + c \cdot a$.

A ring is usually denoted by $(R, +, .)$ and often it is written only as RR when the operations are understood.

For example, the integers Z form a ring, as do the integers modulo n (denoted by $\mathbb{Z}_n$). Less obviously, the square matrices of a given size also form a ring; this ring is noncommutative. Commutative ring theory, or commutative algebra, is much better understood than noncommutative rings are.

As in groups, a ring homomorphism can be defined as a mapping preserving the structure of *both* operations.

Rings are used extensively in algebraic number theory, where "integers" are reimagined as slightly different objects (for example, Gaussian integers), and the effect on concepts such as prime factorization is analyzed. Of particular interest is the fundamental theorem of arithmetic, which involves the concept of unique factorization; in other rings, this may not hold, such as

$$6 = 2 \cdot 3 = \left(1 + \sqrt{-5}\right)\left(1 - \sqrt{-5}\right).$$

Theory developed in this field solves problems ranging from sum of squares theorems to Fermat's last theorem.

Other Applications of Abstract Algebra

Abstract algebra also has heavy application in physics and computer science through the analysis of vector spaces. For example, the Fourier transform and differential geometry both have vector spaces as their underlying structures; in fact, the Poincare conjecture is (roughly speaking) a statement about whether the fundamental group of a manifold determines if the manifold is a sphere.

Related to vector spaces are modules, which are essentially identical to vector spaces but defined over a ring rather than over a field (and are thus more general). Modules are heavily related to representation theory, which views the elements of a group as linear transformations of a vector space; this is desirable to make an abstract object (a group) somewhat more concrete, in the sense that the group is better understood by translating it into a well-understood object in linear algebra (as matrices can be viewed as linear transformations, and vice versa).

The relationships between various algebraic structures are formalized using category theory.

ELEMENTARY ALGEBRA

Elementary algebra encompasses some of the basic concepts of algebra, one of the main branches of mathematics. It is typically taught to secondary school students and builds on their understanding of arithmetic. Whereas arithmetic deals with specified numbers, algebra introduces quantities without fixed values, known as variables. This use of variables entails a use of algebraic notation and an understanding of the general rules of the operators introduced in arithmetic. Unlike abstract algebra, elementary algebra is not concerned with algebraic structures outside the realm of real and complex numbers.

The use of variables to denote quantities allows general relationships between quantities to be formally and concisely expressed, and thus enables solving a broader scope of problems. Many quantitative relationships in science and mathematics are expressed as algebraic equations.

$$x = \frac{-b \pm \sqrt{b^2 - 4ac}}{2a}$$

The quadratic formula, which is the solution to the quadratic equation $ax^2 + bx + c = 0$ where $a \neq 0$. Here the symbols a, b, c represent arbitrary numbers, and x is a variable which represents the solution of the equation.

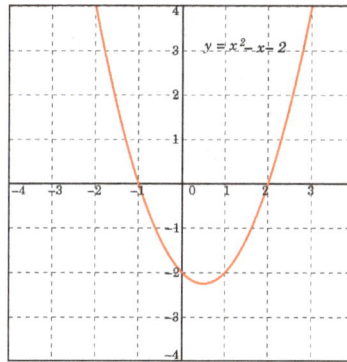

Two-dimensional plot (red curve) of the algebraic equation $y = x^2 - x - 2$.

Algebraic Notation

Algebraic notation describes the rules and conventions for writing mathematical expressions, as well as the terminology used for talking about parts of expressions. For example, the expression $3x^2 - 2xy + c$ has the following components:

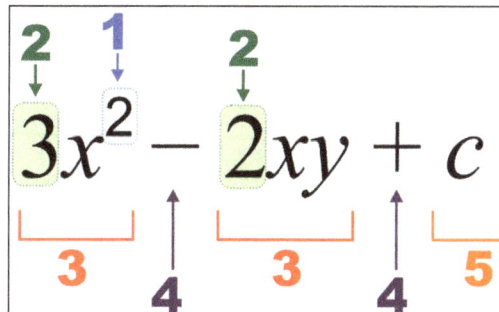

1. Exponent (power),

2. Coefficient,

3. Term,

4. Operator,

5. Constant, *x, y* : variables.

A *coefficient* is a numerical value, or letter representing a numerical constant, that multiplies a variable (the operator is omitted). A *term* is an addend or a summand, a group of coefficients, variables, constants and exponents that may be separated from the other terms by the plus and minus operators. Letters represent variables and constants. By convention, letters at the beginning of the alphabet (e.g. *a,b,c*) are typically used to represent constants, and those toward the end of the alphabet (e.g. *x, y* and *z*) are used to represent variables. They are usually written in italics.

Algebraic operations work in the same way as arithmetic operations, such as addition, subtraction, multiplication, division and exponentiation. and are applied to algebraic variables and terms. Multiplication symbols are usually omitted, and implied when there is no space between two variables or terms, or when a coefficient is used. For example, $3 \times x^2$ is written as $3x^2$, and $2 \times x \times y$ may be written $2xy$.

Usually terms with the highest power (exponent), are written on the left, for example, x^2 is written to the left of x. When a coefficient is one, it is usually omitted (e.g. $1\,x^2$ is written x^2). Likewise when the exponent (power) is one, (e.g. $3x^1$ is written $3x$). When the exponent is zero, the result is always 1 (e.g. x^0 is always rewritten to 1). However 0^0, being undefined, should not appear in an expression, and care should be taken in simplifying expressions in which variables may appear in exponents.

Alternative Notation

Other types of notation are used in algebraic expressions when the required formatting is not available, or can not be implied, such as where only letters and symbols are available. For example, exponents are usually formatted using superscripts, e.g. x^2. In plain text, and in the TeX mark-up language, the caret symbol "^" represents exponents, so x^2 is written as "x^2". In programming languages such as Ada, Fortran, Perl, Python and Ruby, a double asterisk is used, so x^2 is written as "x**2". Many programming languages and calculators use a single asterisk to represent the multiplication symbol, and it must be explicitly used, for example, $3x$ is written "3*x".

Concepts

Variables

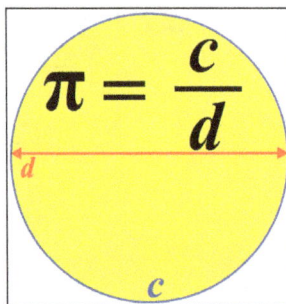

Example of variables showing the relationship between a circle's diameter and its circumference. For any circle, its circumference *c*, divided by its diameter *d*, is equal to the constant pi, π (approximately 3.14).

Elementary algebra builds on and extends arithmetic by introducing letters called variables to represent general (non-specified) numbers. This is useful for several reasons:

1. Variables may represent numbers whose values are not yet known. For example, if the temperature of the current day, C, is 20 degrees higher than the temperature of the previous day, P, then the problem can be described algebraically as $C = P + 20$.

2. Variables allow one to describe general problems, without specifying the values of the quantities that are involved. For example, it can be stated specifically that 5 minutes is equivalent to $60 \times 5 = 300$ seconds. A more general (algebraic) description may state that the number of seconds, $s = 60 \times m$, where m is the number of minutes.

3. Variables allow one to describe mathematical relationships between quantities that may vary. For example, the relationship between the circumference, c, and diameter, d, of a circle is described by $\pi = c / d$.

4. Variables allow one to describe some mathematical properties. For example, a basic property of addition is commutativity which states that the order of numbers being added together does not matter. Commutativity is stated algebraically as $(a + b) = (b + a)$.

Simplifying Expressions

Algebraic expressions may be evaluated and simplified, based on the basic properties of arithmetic operations (addition, subtraction, multiplication, division and exponentiation). For example,

1. Added terms are simplified using coefficients. For example, $x + x + x$ can be simplified as $3x$ (where 3 is a numerical coefficient).

2. Multiplied terms are simplified using exponents. For example, $x \times x \times x$ is represented as x^3.

3. Like terms are added together, for example, $2x^2 + 3ab - x^2 + ab$ is written as $x^2 + 4ab$, because the terms containing x^2 are added together, and, the terms containing ab are added together.

4. Brackets can be "multiplied out", using the distributive property. For example, $x(2x + 3)$ can be written as $(x \times 2x) + (x \times 3)$ which can be written as $2x^2 + 3x$.

5. Expressions can be factored. For example, $6x^5 + 3x^2$, by dividing both terms by $3x^2$ can be written as $3x^2(2x^3 + 1)$.

Equations

An equation states that two expressions are equal using the symbol for equality, = (the equals sign). One of the best-known equations describes Pythagoras' law relating the length of the sides of a right angle triangle:

$$c^2 = a^2 + b^2$$

This equation states that c^2, representing the square of the length of the side that is the hypotenuse (the side opposite the right angle), is equal to the sum (addition) of the squares of the other two sides whose lengths are represented by a and b.

An equation is the claim that two expressions have the same value and are equal. Some equations are true for all values of the involved variables (such as $a + b = b + a$); such equations are called identities. Conditional equations are true for only some values of the involved variables, e.g. $x^2 - 1 = 8$ is true only for $x = 3$ and $x = -3$. The values of the variables which make the equation true are the solutions of the equation and can be found through equation solving.

Another type of equation is an inequality. Inequalities are used to show that one side of the equation is greater, or less, than the other. The symbols used for this are: $a > b$ where $>$ represents 'greater than', and $a < b$ where $<$ represents 'less than'. Just like standard equality equations, numbers can be added, subtracted, multiplied or divided. The only exception is that when multiplying or dividing by a negative number, the inequality symbol must be flipped.

Properties of Equality

By definition, equality is an equivalence relation, meaning it has the properties (a) reflexive (i.e. $b = b$), (b) symmetric (i.e. if $a = b$ then $b = a$) (c) transitive (i.e. if $a = b$ and $b = c$ then $a = c$). It also satisfies the important property that if two symbols are used for equal things, then one symbol can be substituted for the other in any true statement about the first and the statement will remain true. This implies the following properties:

- If $a = b$ and $c = d$ then $a + c = b + d$ and $ac = bd$;
- If $a = b$ then $a + c = b + c$ and $ac = bc$;
- More generally, for any function f, if $a = b$ then $f(a) = f(b)$.

Properties of Inequality

The relations *less than* $<$ and $>$ greater than have the property of transitivity:

- If $a < b$ and $b < c$ then $a < c$;
- If $a < b$ and $c < d$ then $a + c < b + d$;
- If $a < b$ and $c > 0$ then $ac < bc$;
- If $a < b$ and $c < 0$ then $bc < ac$.

By reversing the inequation, $<$ and $>$ can be swapped, for example:

- $a < b$ is equivalent to $b > a$.

Substitution

Substitution is replacing the terms in an expression to create a new expression. Substituting 3 for a in the expression $a*5$ makes a new expression $3*5$ with meaning 15. Substituting the terms of a statement makes a new statement. When the original statement is true independently of the values of the terms, the statement created by substitutions is also true. Hence definitions can be made in symbolic terms and interpreted through substitution: if $a^2 := a \times a$ is meant as the definition of a^2 as the product of a with itself, substituting 3 for a informs the reader of this statement that 3^2 means $3 \times 3 = 9$. Often it's not known whether the statement is true independently of the values of the terms. And, substitution allows one to derive restrictions on the possible values, or show what conditions the statement holds under. For example, taking the statement $x + 1 = 0$, if x is substituted with 1, this implies $1 + 1 = 2 = 0$, which is false, which implies that if $x + 1 = 0$ then x canot be 1.

If x and y are integers, rationals, or real numbers, then $xy = 0$ implies $x = 0$ or $y = 0$. Consider $abc = 0$. Then, substituting a for x and bc for y, we learn $a = 0$ or $bc = 0$. Then we can substitute again, letting $x = b$ and $y = c$, to show that if $bc = 0$ then $b = 0$ or $c = 0$. Therefore, if $abc = 0$, then $a = 0$ or ($b = 0$ or $c = 0$), so $abc = 0$ implies $a = 0$ or $b = 0$ or $c = 0$.

If the original fact were stated as "$ab = 0$ implies $a = 0$ or $b = 0$", then when saying "consider $abc = 0$," we would have a conflict of terms when substituting. Yet the above logic is still valid to show that if $abc = 0$ then $a = 0$ or $b = 0$ or $c = 0$ if, instead of letting $a = a$ and $b = bc$, one substitutes a for a and b for bc (and with $bc = 0$, substituting b for a and c for b). This shows that substituting for the terms in a statement isn't always the same as letting the terms from the statement equal the substituted terms. In this situation it's clear that if we substitute an expression a into the a term of the original equation, the a substituted does not refer to the a in the statement "$ab = 0$ implies $a = 0$ or $b = 0$."

Solving Algebraic Equations

A typical algebra problem.

The following sections lay out examples of some of the types of algebraic equations that may be encountered:

Linear Equations with One Variable

Linear equations are so-called, because when they are plotted, they describe a straight line. The simplest equations to solve are linear equations that have only one variable. They contain only constant numbers and a single variable without an exponent. As an example, consider:

> Problem in words: If you double the age of a child and add 4, the resulting answer is 12. How old is the child?

> Equivalent equation: $2x + 4 = 12$ where x represent the child's age.

To solve this kind of equation, the technique is add, subtract, multiply, or divide both sides of the equation by the same number in order to isolate the variable on one side of the equation. Once the variable is isolated, the other side of the equation is the value of the variable. This problem and its solution are as follows:

1. Equation to solve:	$2x + 4 = 12$
2. Subtract 4 from both sides:	$2x + 4 - 4 = 12 - 4$
3. This simplifies to:	$2x = 8$
4. Divide both sides by 2:	$\dfrac{2x}{2} = \dfrac{8}{2}$
5. This simplifies to the solution:	$x = 4$

In words: the child is 4 years old.

The general form of a linear equation with one variable, can be written as: $ax + b = c$.

Following the same procedure (i.e. subtract b from both sides, and then divide by a), the general solution is given by $x = \dfrac{c - b}{a}$.

Linear Equations with Two Variables

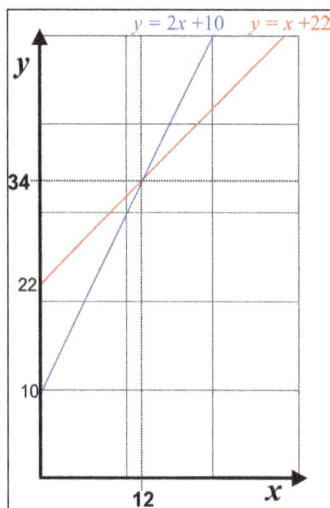

Solving two linear equations with a unique solution
at the point that they intersect.

A linear equation with two variables has many (i.e. an infinite number of) solutions. For example:

Problem in words: A father is 22 years older than his son. How old are they?

Equivalent equation: $y = x + 22$ where y is the father's age, x is the son's age.

This cannot be worked out by itself. If the son's age was made known, then there would no longer be two unknowns (variables), and the problem becomes a linear equation with just one variable, that can be solved as described above.

To solve a linear equation with two variables (unknowns), requires two related equations. For example, if it was also revealed that:

Problem in Words

In 10 years, the father will be twice as old as his son.

Equivalent Equation

$$y+10 = 2\times(x+10)$$
$$y = 2\times(x+10)-10 \quad \text{Subtract 10 from both sides}$$
$$y = 2x+20-10 \quad\quad \text{Multiple out brackets}$$
$$y = 2x+10 \quad\quad\quad \text{Simplify}$$

Now there are two related linear equations, each with two unknowns, which enables the production of a linear equation with just one variable, by subtracting one from the other (called the elimination method):

$$\begin{cases} y = x+22 & \text{First equation} \\ y = 2x+10 & \text{Second equation} \end{cases}$$

$$\quad\quad\quad\quad\quad\quad\quad\quad \text{Subtract the first equation from}$$
$$(y-y) = (2x-x)+10-22 \quad \text{the second in order to remove } y$$
$$0 = x-12 \quad\quad\quad\quad \text{Simplify}$$
$$12 = x \quad\quad\quad\quad\quad \text{Add 12 to both sides}$$
$$x = 12 \quad\quad\quad\quad\quad \text{Rearrange}$$

In other words, the son is aged 12, and since the father 22 years older, he must be 34. In 10 years time, the son will be 22, and the father will be twice his age, 44. This problem is illustrated on the associated plot of the equations.

Quadratic Equations

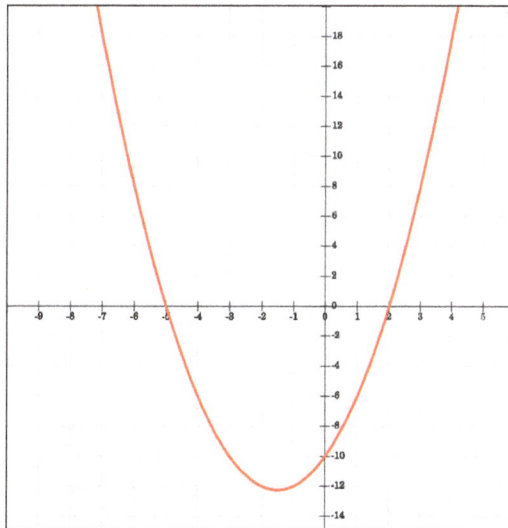

Quadratic equation plot of $y = x^2+3x-10$ showing its roots at $x = -5$ and $x = 2$, and that the quadratic can be rewritten as $y = (x+5)(x-2)$.

A quadratic equation is one which includes a term with an exponent of 2, for example, x^2, and no term with higher exponent. The name derives from the Latin *quadrus*, meaning square. In general, a quadratic equation can be expressed in the form $ax^2 + bx + c = 0$, where a is not zero (if it were zero, then the equation would not be quadratic but linear). Because of this a quadratic equation must contain the term ax^2, which is known as the quadratic term. Hence $a \neq 0$, and so we may divide by a and rearrange the equation into the standard form

$$x^2 + px + q = 0$$

where $p = \dfrac{b}{a}$ and $q = \dfrac{c}{a}$. Solving this, by a process known as completing the square, leads to the quadratic formula

$$x = \frac{-b \pm \sqrt{b^2 - 4ac}}{2a},$$

where the symbol "$\pm$" indicates that both

$$x = \frac{-b + \sqrt{b^2 - 4ac}}{2a} \quad \text{and} \quad x = \frac{-b - \sqrt{b^2 - 4ac}}{2a}$$

are solutions of the quadratic equation.

Quadratic equations can also be solved using factorization (the reverse process of which is expansion, but for two linear terms is sometimes denoted foiling). As an example of factoring:

$$x^2 + 3x - 10 = 0,$$

which is the same thing as,

$$(x + 5)(x - 2) = 0.$$

It follows from the zero-product property that either $x = 2$ or $x = -5$ are the solutions, since precisely one of the factors must be equal to zero. All quadratic equations will have two solutions in the complex number system, but need not have any in the real number system. For example,

$$x^2 + 1 = 0$$

has no real number solution since no real number squared equals −1. Sometimes a quadratic equation has a root of multiplicity 2, such as:

$$(x + 1)^2 = 0.$$

For this equation, −1 is a root of multiplicity 2. This means −1 appears twice, since the equation can be rewritten in factored form as,

$$[x - (-1)][x - (-1)] = 0.$$

Complex Numbers

All quadratic equations have exactly two solutions in complex numbers (but they may be equal to each other), a category that includes real numbers, imaginary numbers, and sums of real and imaginary numbers. Complex numbers first arise in the teaching of quadratic equations and the quadratic formula. For example, the quadratic equation:

$$x^2 + x + 1 = 0$$

has solutions

$$x = \frac{-1+\sqrt{-3}}{2} \quad \text{and} \quad x = \frac{-1-\sqrt{-3}}{2}.$$

Since $\sqrt{-3}$ is not any real number, both of these solutions for x are complex numbers.

Exponential and Logarithmic Equations

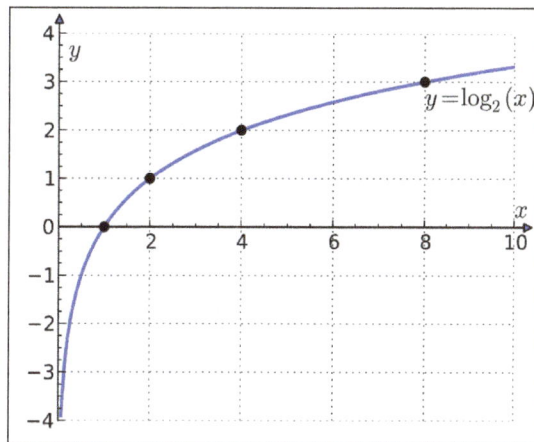

The graph of the logarithm to base 2 crosses the x axis (horizontal axis) at 1 and passes through the points with coordinates (2, 1), (4, 2), and (8, 3). For example, $\log_2(8) = 3$, because $2^3 = 8$. The graph gets arbitrarily close to the y axis, but does not meet or intersect t.

An exponential equation is one which has the form $a^x = b$ for $a > 0$. which has solution

$$X = \log_a b = \frac{\ln b}{\ln a}$$

when $b > 0$. Elementary algebraic techniques are used to rewrite a given equation in the above way before arriving at the solution. For example, if

$$3 \cdot 2^{x-1} + 1 = 10$$

then, by subtracting 1 from both sides of the equation, and then dividing both sides by 3 we obtain

$$2^{x-1} = 3$$

whence

$$x - 1 = \log_2 3$$

or

$$x = \log_2 3 + 1.$$

A logarithmic equation is an equation of the form $log_a(x) = b$ for $a > 0$, which has solution

$$X = a^b.$$

For example, if

$$4\log_5(x-3) - 2 = 6$$

then, by adding 2 to both sides of the equation, followed by dividing both sides by 4, we get

$$\log_5(x-3) = 2$$

whence

$$x - 3 = 5^2 = 25$$

from which we obtain

$$x = 28.$$

Radical Equations

$$\sqrt[2]{x^3} \equiv x^{\frac{3}{2}}$$

Radical equation showing two ways to represent the same expression. The triple bar means the equation is true for all values of x.

A radical equation is one that includes a radical sign, which includes square roots, $\sqrt{x}$ cube roots, $\sqrt[3]{x}$, and nth roots, $\sqrt[n]{x}$ Recall that an nth root can be rewritten in exponential format, so that $\sqrt[n]{x}$ is equivalent to $x^{\frac{1}{n}}$. Combined with regular exponents (powers), then $\sqrt[2]{x^3}$ (the square root of x cubed), can be rewritten as $x^{\frac{3}{2}}$. So a common form of a radical equation is $\sqrt[n]{x^m} = a$ (equivalent to $x^{\frac{m}{n}} = a$) where m and n are integers. It has real solution(s):

n is odd	n is even and $a \geq 0$	n and m are even and $a < 0$	n is even, m is odd, and $a < 0$
			no real solution
$x = \sqrt[n]{a^m}$ equivalently $x = (\sqrt[n]{a})^m$	$x = \pm\sqrt[n]{a^m}$ equivalently $x = \pm(\sqrt[n]{a})^m$	$x = \pm\sqrt[n]{a^m}$	

For example, if:

$$(x+5)^{2/3} = 4$$

then

$$x + 5 = \pm(\sqrt{4})^3,$$
$$x + 5 = \pm 8,$$
$$x = -5 \pm 8,$$

and thus

$$x = 3 \quad \text{or} \quad x = -13$$

System of Linear Equations

There are different methods to solve a system of linear equations with two variables.

Elimination Method

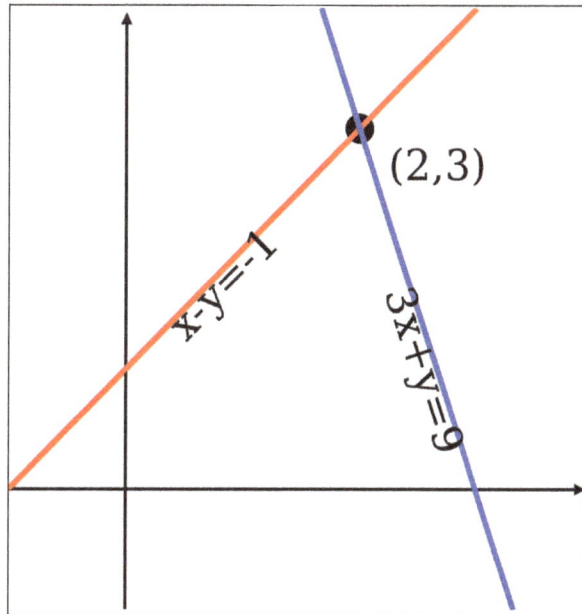

The solution set for the equations $x - y = -1$ and $3x + y = 9$ is the single point (2, 3).

An example of solving a system of linear equations is by using the elimination method:

$$\begin{cases} 4x + 2y = 14 \\ 2x - y = 1. \end{cases}$$

Multiplying the terms in the second equation by 2:

$$4x + 2y = 14$$
$$4x - 2y = 2.$$

Adding the two equations together to get:

$$8x = 16$$

which simplifies to

$$x = 2.$$

Since the fact that $x = 2$ is known, it is then possible to deduce that $y = 3$ by either of the original two equations (by using 2 instead of x) The full solution to this problem is then.

$$\begin{cases} x = 2 \\ y = 3. \end{cases}$$

This is not the only way to solve this specific system; y could have been solved before x.

Substitution Method

Another way of solving the same system of linear equations is by substitution.

$$\begin{cases} 4x + 2y & = 14 \\ 2x - y & = 1. \end{cases}$$

An equivalent for y can be deduced by using one of the two equations. Using the second equation:

$$2x - y = 1$$

Subtracting $2x$ from each side of the equation:

$$\begin{aligned} 2x - 2x - y & = 1 - 2x \\ -y & = 1 - 2x \end{aligned}$$

and multiplying by −1:

$$y = 2x - 1.$$

Using this y value in the first equation in the original system:

$$\begin{aligned} 4x + 2(2x - 1) & = 14 \\ 4x + 4x - 2 & = 14 \\ 8x - 2 & = 14 \end{aligned}$$

Adding 2 on each side of the equation:

$$\begin{aligned} 8x - 2 + 2 & = 14 + 2 \\ 8x & = 16 \end{aligned}$$

which simplifies to

$$x = 2$$

Using this value in one of the equations, the same solution as in the previous method is obtained.

$$\begin{cases} x = 2 \\ y = 3. \end{cases}$$

This is not the only way to solve this specific system; in this case as well, y could have been solved before x.

Other Types of Systems of Linear Equations

Inconsistent Systems

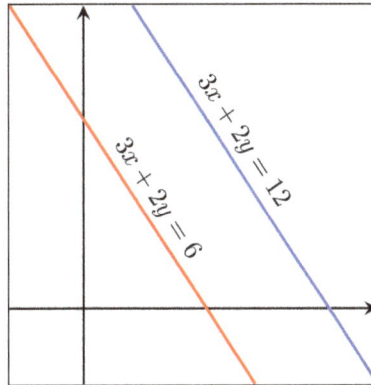

The equations $3x + 2y = 6$ and $3x + 2y = 12$ are parallel
and cannot intersect, and is unsolvable.

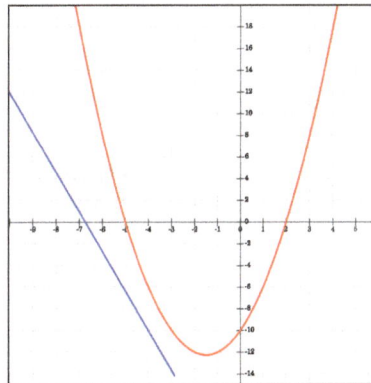

Plot of a quadratic equation (red) and a linear equation (blue)
that do not intersect, and consequently for which there is no common solution.

In the above example, a solution exists. However, there are also systems of equations which do not have any solution. Such a system is called inconsistent. An obvious example is,

$$\begin{cases} x + y & = 1 \\ 0x + 0y & = 2. \end{cases}$$

As $0 \neq 2$, the second equation in the system has no solution. Therefore, the system has no solution. However, not all inconsistent systems are recognized at first sight. As an example, consider the system.

$$\begin{cases} 4x + 2y & = 12 \\ -2x - y & = -4. \end{cases}$$

Multiplying by 2 both sides of the second equation, and adding it to the first one results in

$$0x + 0y = 4,$$

which clearly has no solution.

Undetermined Systems

There are also systems which have infinitely many solutions, in contrast to a system with a unique solution (meaning, a unique pair of values for x and y) For example:

$$\begin{cases} 4x + 2y = 12 \\ -2x - y = -6 \end{cases}$$

Isolating y in the second equation:

$$y = -2x + 6$$

And using this value in the first equation in the system:

$$4x + 2(-2x + 6) = 12$$
$$4x - 4x + 12 = 12$$
$$12 = 12$$

The equality is true, but it does not provide a value for x. Indeed, one can easily verify (by just filling in some values of x) that for any x there is a solution as long as $y = -2x + 6$. There is an infinite number of solutions for this system.

Over- and Underdetermined Systems

Systems with more variables than the number of linear equations are called underdetermined. Such a system, if it has any solutions, does not have a unique one but rather an infinitude of them. An example of such a system is,

$$\begin{cases} x + 2y = 10 \\ y - z = 2. \end{cases}$$

When trying to solve it, one is led to express some variables as functions of the other ones if any solutions exist, but cannot express *all* solutions numerically because there are an infinite number of them if there are any.

A system with a greater number of equations than variables is called overdetermined. If an overdetermined system has any solutions, necessarily some equations are linear combinations of the others.

LINEAR ALGEBRA

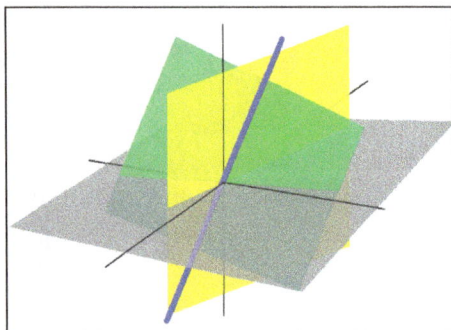

In the three-dimensional Euclidean space, these three planes represent solutions of linear equations and their intersection represents the set of common solutions: in this case, a unique point. The blue line is the common solution of a pair of linear equations.

Linear algebra is the branch of mathematics concerning linear equations such as

$$a_1x_1 + \cdots + a_nx_n = b,$$

linear functions such as

$$(x_1, \ldots, x_n) \mapsto a_1x_1 + \ldots + a_nx_n,$$

and their representations through matrices and vector spaces.

Linear algebra is central to almost all areas of mathematics. For instance, linear algebra is fundamental in modern presentations of geometry, including for defining basic objects such as lines, planes and rotations. Also, functional analysis may be basically viewed as the application of linear algebra to spaces of functions. Linear algebra is also used in most sciences and engineering areas, because it allows modeling many natural phenomena, and efficiently computing with such models. For nonlinear systems, which cannot be modeled with linear algebra, linear algebra is often used as a first-order approximation.

Vector Spaces

Until the 19th century, linear algebra was introduced through systems of linear equations and matrices. In modern mathematics, the presentation through *vector spaces* is generally preferred, since it is more synthetic, more general (not limited to the finite-dimensional case), and conceptually simpler, although more abstract.

A vector space over a field F (often the field of the real numbers) is a set V equipped with two binary operations satisfying the following axioms. Elements of V are called *vectors*, and elements of F are called *scalars*. The first operation, *vector addition*, takes any two vectors v and w and outputs a third vector $v + w$. The second operation, *scalar multiplication*, takes any scalar a and any vector v and outputs a new vector av. The axioms that addition and scalar multiplication must satisfy are the following. (In the list below, u, v and w are arbitrary elements of V, and a and b are arbitrary scalars in the field F.)

Axiom	Signification
Associativity of addition	$u + (v + w) = (u + v) + w$
Commutativity of addition	$u + v = v + u$
Identity element of addition	There exists an element o in V, called the *zero vector* (or simply *zero*), such that $v + o = v$ for all v in V.
Inverse elements of addition	For every v in V, there exists an element $-v$ in V, called the *additive inverse* of v, such that $v + (-v) = o$
Distributivity of scalar multiplication with respect to vector addition	$a(u + v) = au + av$
Distributivity of scalar multiplication with respect to field addition	$(a + b)v = av + bv$

Compatibility of scalar multiplication with field multiplication	$a(bv) = (ab)v$
Identity element of scalar multiplication	$1v = v$, where 1 denotes the multiplicative identity of F.

The first four axioms mean that V is an abelian group under addition.

Elements of a vector space may have various nature; for example, they can be sequences, functions, polynomials or matrices. Linear algebra is concerned with properties common to all vector spaces.

Linear Maps

Linear maps are mappings between vector spaces that preserve the vector-space structure. Given two vector spaces V and W over a field F, a linear map (also called, in some contexts, linear transformation, linear mapping or linear operator) is a map,

$$T\ V \to W$$

that is compatible with addition and scalar multiplication, that is

$$T(u+v) = T(u) + T(v), \quad T(av) = aT(v)$$

for any vectors u,v in V and scalar a in F.

This implies that for any vectors u, v in V and scalars a, b in F, one has

$$T(au+bv) = T(au) + T(bv) = aT(u) + bT(v)$$

When a bijective linear map exists between two vector spaces (that is, every vector from the second space is associated with exactly one in the first), the two spaces are isomorphic. Because an isomorphism preserves linear structure, two isomorphic vector spaces are "essentially the same" from the linear algebra point of view, in the sense that they cannot be distinguished by using vector space properties. An essential question in linear algebra is testing whether a linear map is an isomorphism or not, and, if it is not an isomorphism, finding its range (or image) and the set of elements that are mapped to the zero vector, called the kernel of the map. All these questions can be solved by using Gaussian elimination or some variant of this algorithm.

Subspaces, Span and Basis

The study of subsets of vector spaces that are themselves vector spaces for the induced operations is fundamental, similarly as for many mathematical structures. These subsets are called linear subspaces. More precisely, a linear subspace of a vector space V over a field F is a subset W of V such that $u + v$ and au are in W, for every u, v in W, and every a in F. (These conditions suffice for implying that W is a vector space).

For example, given a linear map $T : V \to W$, the image $T(V)$ of V, and the inverse image $T^{-1}(0)$ of 0 (called kernel or null space), are linear subspaces of W and V, respectively.

Another important way of forming a subspace is to consider linear combinations of a set S of vectors: the set of all sums

$$a_1v_1 + a_2v_2 + \cdots + a_kv_k,$$

where $v_1, v_2, ..., v_k$ are in V, and $a_1, a_2, ..., a_k$ are in F form a linear subspace called the span of S. The span of S is also the intersection of all linear subspaces containing S. In other words, it is the (smallest for the inclusion relation) linear subspace containing S.

A set of vectors is linearly independent if none is in the span of the others. Equivalently, a set S of vector is linearly independent if the only way to express the zero vector as a linear combination of elements of S is to take zero for every coefficient a_i.

A set of vectors that spans a vector space is called a spanning set or generating set. If a spanning set S is *linearly dependent* (that is not linearly independent), then some element w of S is in the span of the other elements of S, and the span would remain the same if one remove w from S. One may continue to remove elements of S until getting a *linearly independent spanning set*. Such a linearly independent set that spans a vector space V is called a basis of V. The importance of bases lies in the fact that there are together minimal generating sets and maximal independent sets. More precisely, if S is a linearly independent set, and T is a spanning set such that $S \subset T$, then there is a basis B such that $S \subseteq B \subseteq T$.

Any two bases of a vector space V have the same cardinality, which is called the dimension of V; this is the dimension theorem for vector spaces. Moreover, two vector spaces over the same field F are isomorphic if and only if they have the same dimension.

If any basis of V (and therefore every basis) has a finite number of elements, V is a *finite-dimensional vector space*. If U is a subspace of V, then $\dim U \le \dim V$. In the case where V is finite-dimensional, the equality of the dimensions implies $U = V$.

If U_1 and U_2 are subspaces of V, then

$$\dim(U_1 + U_2) = \dim U_1 + \dim U_2 - \dim(U_1 \cap U_2),$$

where $U_1 + U_2$ denotes the span of $U_1 \cup U_2$.

Matrices

Matrices allow explicit manipulation of finite-dimensional vector spaces and linear maps. Their theory is thus an essential part of linear algebra.

Let V be a finite-dimensional vector space over a field F, and $(v_1, v_2, ..., v_m)$ be a basis of V (thus m is the dimension of V). By definition of a basis, the map

$$(a_1, ..., a_m) \mapsto a_1v_1 + \cdots a_mv_m$$
$$F^m \to V$$

is a bijection from F^m, the set of the sequences of m elements of F, onto V. This is an isomorphism of vector spaces, if F^m, is equipped of its standard structure of vector space, where vector addition and scalar multiplication are done component by component.

This isomorphism allows representing a vector by its inverse image under this isomorphism, that is by the coordinates vector $(a_1, \ldots, a_m)$ or by the column matrix

$$\begin{bmatrix} a_1 \\ \vdots \\ a_m \end{bmatrix}.$$

If W is another finite dimensional vector space (possibly the same), with a basis $(w, \ldots, w)$, a linear map f from W to V is well defined by its values on the basis elements, that is $(f(w_1), \ldots, f(w_n))$. Thus, f is well represented by the list of the corresponding column matrices. That is, if

$$f(w_j) = a_{1,j} v_1 + \cdots + a_{m,j} v_m,$$

for $j = 1, \ldots, n$, then f is represented by the matrix

$$\begin{bmatrix} a_{1,1} & \cdots & a_{1,n} \\ \vdots & \cdots & \vdots \\ a_{m,1} & \cdots & a_{m,n} \end{bmatrix},$$

with m rows and n columns.

Matrix multiplication is defined in such a way that the product of two matrices is the matrix of the composition of the corresponding linear maps, and the product of a matrix and a column matrix is the column matrix representing the result of applying the represented linear map to the represented vector. It follows that the theory of finite-dimensional vector spaces and the theory of matrices are two different languages for expressing exactly the same concepts.

Two matrices that encode the same linear transformation in different bases are called similar. Equivalently, two matrices are similar if one can transform one in the other by elementary row and column operations. For a matrix representing a linear map from W to V, the row operations correspond to change of bases in V and the column operations correspond to change of bases in W. Every matrix is similar to an identity matrix possibly bordered by zero rows and zero columns. In terms of vector space, this means that, for any linear map from W to V, there are bases such that a part of the basis of W is mapped bijectively on a part of the basis of V, and that the remaining basis elements of W, if any, are mapped to zero (this is a way of expressing the fundamental theorem of linear algebra). Gaussian elimination is the basic algorithm for finding these elementary operations, and proving this theorem.

Linear Systems

A finite set of linear equations in a finite set of variables, for example, $x_1, x_2, \ldots, x_n$ or $x, y, \ldots, z$ is called a system of linear equations or a linear system.

Systems of linear equations form a fundamental part of linear algebra. Historically, linear algebra and matrix theory has been developed for solving such systems. In the modern presentation of linear algebra through vector spaces and matrices, many problems may be interpreted in terms of linear systems.

For example, let:

$$2x + y - z = 8$$
$$-3x - y + 2z = -11$$
$$-2x + y + 2z = -3$$

be a linear system.

To such a system, one may associate its matrix:

$$M \begin{bmatrix} 2 & 1 & -1 \\ -3 & -1 & 2 \\ -2 & 1 & 2 \end{bmatrix}.$$

and its right member vector:

$$v = \begin{bmatrix} 8 \\ -11 \\ -3 \end{bmatrix}.$$

Let T be the linear transformation associated to the matrix M. A solution of the system (S) is a vector:

$$X = \begin{bmatrix} x \\ y \\ z \end{bmatrix}$$

such that,

$$T(X) = v,$$

that is an element of the preimage of v by T.

Let (S') be the associated homogeneous system, where the right-hand sides of the equations are put to zero:

$$2x + y - z = 0$$
$$-3x - y + 2z = 0$$
$$-2x + y + 2z = 0$$

The solutions of (S') are exactly the elements of the kernel of T or, equivalently, M.

The Gaussian-elimination consists of performing elementary row operations on the augmented matrix:

$$M \begin{bmatrix} 2 & 1 & -1 & 8 \\ -3 & -1 & 2 & -11 \\ -2 & 1 & 2 & -3 \end{bmatrix}$$

for putting it in reduced row echelon form. These row operations do not change the set of solutions of the system of equations. In the example, the reduced echelon form is:

$$M \begin{bmatrix} 1 & 0 & 0 & 2 \\ 0 & 1 & 0 & 3 \\ 0 & 0 & 1 & -1 \end{bmatrix},$$

showing that the system (S) has the unique solution:

$$x = 2$$
$$y = 3$$
$$z = -1.$$

It follows from this matrix interpretation of linear systems that the same methods can be applied for solving linear systems and for many operations on matrices and linear transformations, which include the computation of the ranks, kernels, matrix inverses.

Endomorphisms and Square Matrices

A linear endomorphism is a linear map that maps a vector space V to itself. If V has a basis of n elements, such an endomorphism is represented by a square matrix of size n.

With respect to general linear maps, linear endomorphisms and square matrices have some specific properties that make their study an important part of linear algebra, which is used in many parts of mathematics, including geometric transformations, coordinate changes, quadratic forms, and many other part of mathematics.

Determinant

The *determinant* of a square matrix A is de ined to be:

$$\sum_{\sigma \in S_n} (-1)^{\sigma} a_{1\sigma(1)} \cdots a_{n\sigma(n)},$$

where S_n is the group of all permutations of n elements, σ is a permutation, and $(-1)^{\sigma}$ the parity of the permutation. A matrix is invertible if and only if the determinant is invertible (i.e., nonzero if the scalars belong to a field).

Cramer's rule is a closed-form expression, in terms of determinants, of the solution of a system of n linear equations in n unknowns. Cramer's rule is useful for reasoning about the solution, but, except for $n = 2$ or 3, it is rarely used for computing a solution, since Gaussian elimination is a faster algorithm.

The *determinant of an endomorphism* is the determinant of the matrix representing the endomorphism in terms of some ordered basis. This definition makes sense, since this determinant is independent of the choice of the basis.

Eigenvalues and Eigenvectors

If f is a linear endomorphism of a vector space V over a field F, an *eigenvector* of f is a nonzero vector v of V such that $f(v) = av$ for some scalar a in F. This scalar a is an *eigenvalue* of f.

If the dimension of V is finite, and a basis has been chosen, f and v may be represented, respectively, by a square matrix M and a column matrix z; the equation defining eigenvectors and eigenvalues becomes:

$$Mz = az.$$

Using the identity matrix I, whose entries are all zero, except those of the main diagonal, which are equal to one, this may be rewritten:

$$(M - aI)z = 0.$$

As z is supposed to be nonzero, this means that $M - aI$ is a singular matrix, and thus that its determinant $\det(M - aI)$ equals zero. The eigenvalues are thus the roots of the polynomial:

$$\det(xI - M).$$

If V is of dimension n, this is a monic polynomial of degree n, called the characteristic polynomial of the matrix (or of the endomorphism), and there are, at most, n eigenvalues.

If a basis exists that consists only of eigenvectors, the matrix of f on this basis has a very simple structure: it is a diagonal matrix such that the entries on the main diagonal are eigenvalues, and the other entries are zero. In this case, the endomorphism and the matrix are said diagonalizable. More generally, an endomorphism and a matrix are also said diagonalizable, if they become diagonalizable after extending the field of scalars. In this extended sense, if the characteristic polynomial is square-free, then the matrix is diagonalizable.

A symmetric matrix is always diagonalizable. There are non-diagonalizable matrices, the simplest being,

$$\begin{bmatrix} 0 & 1 \\ 0 & 0 \end{bmatrix}$$

(it cannot be diagonalizable since its square is the zero matrix, and the square of a nonzero diagonal matrix is never zero).

When an endomorphism is not diagonalizable, there are bases on which it has a simple form, although not as simple as the diagonal form. The Frobenius normal form does not need of extending the field of scalars and makes the characteristic polynomial immediately readable on the matrix. The Jordan normal form requires to extend the field of scalar for containing all eigenvalues, and differs from the diagonal form only by some entries that are just above the main diagonal and are equal to 1.

Duality

A linear form is a linear map from a vector space V over a field F to the field of scalars F, viewed as a vector space over itself. Equipped by pointwise addition and multiplication by a scalar, the linear forms form a vector space, called the dual space of V, and usually denoted V^*.

If $v_1, \ldots, v_n$ is a basis of V (this implies that V is finite-dimensional), then one can define, for $i = 1, \ldots, n$, a linear map v_i^* such that $v_i^*(e_i) = 1$ and $v_i^*(e_j) = 0$ if $j \neq i$. These linear maps form a basis of

V^* called the dual basis of $v_1,\ldots,v_n$. (If V is not finite-dimensional, the v_i^* may be defined similarly; they are linearly independent, but do not form a basis.)

For v in V, the map

$$f \to f(v)$$

is a linear form on V^* This defines the canonical linear map from V into V^{**} the dual of called the bidual of V^*. This canonical map is an isomorphism if V is finite-dimensional, and this allows identifying V with its bidual. (In the infinite dimensional case, the canonical map is injective, but not surjective).

There is thus a complete symmetry between a finite-dimensional vector space and its dual. This motivates the frequent use, in this context, of the bra–ket notation

$$\langle f,x \rangle$$

for denoting $f(x)$.

Dual Map

Let

$$f:V \to W$$

be a linear map. For every linear form h on W, the composite function $h \circ f$ is a linear form on V. This defines a linear map

$$f^*:W^* \to V^*$$

between the dual spaces, which is called the dual or the transpose of f.

If V and W are finite dimensional, and M is the matrix of f in terms of some ordered bases, then the matrix of f^* over the dual bases is the transpose M^T of M, obtained by exchanging rows and columns.

If elements of vector spaces and their duals are represented by column vectors, this duality may be expressed in bra–ket notation by

$$\langle h^\mathsf{T}, Mv \rangle = \langle h^\mathsf{T}M, v \rangle.$$

For highlighting this symmetry, the two members of this equality are sometimes written

$$\langle h^\mathsf{T} \,|\, M \,|\, v \rangle.$$

Inner-product Spaces

Besides these basic concepts, linear algebra also studies vector spaces with additional structure, such as an inner product. The inner product is an example of a bilinear form, and it gives the vector space a geometric structure by allowing for the definition of length and angles. Formally, an *inner product* is a map

$$\langle \cdot,\cdot \rangle : V \times V \to F$$

that satisfies the following three axioms for all vectors u, v, w in V and all scalars a in F:

- Conjugate symmetry:

$$\langle u,v \rangle = \overline{\langle v,u \rangle}.$$

In R, it is symmetric.

- Linearity in the first argument:

$$\langle au,v \rangle = a\langle u,v \rangle.$$
$$\langle u+v,w \rangle = \langle u,w \rangle + \langle v,w \rangle.$$

- Positive-definiteness:

$$\langle v,v \rangle \geq 0 \text{ with equality only for } v = o.$$

We can define the length of a vector v in V by

$$\| v \|^2 = \langle v,v \rangle,$$

and we can prove the Cauchy–Schwarz inequality:

$$| \langle u,v \rangle | \leq \| u \| \| v \|.$$

In particular, the quantity

$$\frac{| \langle u,v \rangle |}{\| u \| \| v \|} \leq 1,$$

and so we can call this quantity the cosine of the angle between the two vectors.

Two vectors are orthogonal if $\langle u,v \rangle = 0$. An orthonormal basis is a basis where all basis vectors have length 1 and are orthogonal to each other. Given any finite-dimensional vector space, an orthonormal basis could be found by the Gram–Schmidt procedure. Orthonormal bases are particularly easy to deal with, since if $v = a_1 v_1 + \ldots + a_n v_n$, then $a_i = \langle v,v_i \rangle$.

The inner product facilitates the construction of many useful concepts. For instance, given a transform T, we can define its Hermitian conjugate T^* as the linear transform satisfying

$$\langle Tu,v \rangle = \langle u,T^*v \rangle.$$

If T satisfies $TT^* = T^*T$, we call T normal. It turns out that normal matrices are precisely the matrices that have an orthonormal system of eigenvectors that span V.

Relationship with Geometry

There is a strong relationship between linear algebra and geometry, which started with the introduction by René Descartes, in 1637, of Cartesian coordinates. In this new (at that time) geometry, now called Cartesian geometry, points are represented by Cartesian coordinates, which are sequences of three real numbers (in the case of the usual three-dimensional space). The basic objects of geometry, which are lines and planes are represented by linear equations. Thus, computing intersections of lines and planes amounts to solving systems of linear equations. This was one of the main motivations for developing linear algebra.

Most geometric transformation, such as translations, rotations, reflections, rigid motions, isometries, and projections transform lines into lines. It follows that they can be defined, specified and studied in terms of linear maps. This is also the case of homographies and Möbius transformations, when considered as transformations of a projective space.

Until the end of 19th century, geometric spaces were defined by axioms relating points, lines and planes (synthetic geometry). Around this date, it appeared that one may also define geometric spaces by constructions involving vector spaces. It has been shown that the two approaches are essentially equivalent. In classical geometry, the involved vector spaces are vector spaces over the reals, but the constructions may be extended to vector spaces over any field, allowing considering geometry over arbitrary fields, including finite fields.

Presently, most textbooks, introduce geometric spaces from linear algebra, and geometry is often presented, at elementary level, as a subfield of linear algebra.

Usage and Applications

Linear algebra is used in almost all areas of mathematics, thus making it relevant in almost all scientific domains that use mathematics. These applications may be divided into several wide categories.

Geometry of our Ambient Space

The modeling of our ambient space is based on geometry. Sciences concerned with this space use geometry widely. This is the case with mechanics and robotics, for describing rigid body dynamics; geodesy for describing Earth shape; perspectivity, computer vision, and computer graphics, for describing the relationship between a scene and its plane representation; and many other scientific domains.

In all these applications, synthetic geometry is often used for general descriptions and a qualitative approach, but for the study of explicit situations, one must compute with coordinates. This requires the heavy use of linear algebra.

Functional Analysis

Functional analysis studies function spaces. These are vector spaces with additional structure, such as Hilbert spaces. Linear algebra is thus a fundamental part of functional analysis and its applications, which include, in particular, quantum mechanics (wave functions).

Study of Complex Systems

Most physical phenomena are modeled by partial differential equations. To solve them, one usually decomposes the space in which the solutions are searched into small, mutually interacting cells. For linear systems this interaction involves linear functions. For nonlinear systems, this interaction is often approximated by linear functions. In both cases, very large matrices are generally involved. Weather forecasting is a typical example, where the whole Earth atmosphere is divided in cells of, say, 100 km of width and 100 m of height.

Scientific Computation

Nearly all scientific computations involve linear algebra. Consequently, linear algebra algorithms have been highly optimized. BLAS and LAPACK are the best known implementations. For improving efficiency, some of them configure the algorithms automatically, at run time, for adapting them to the specificities of the computer (cache size, number of available cores, etc.).

Some processors, typically graphics processing units (GPU), are designed with a matrix structure, for optimizing the operations of linear algebra.

Extensions and Generalizations

Module Theory

The existence of multiplicative inverses in fields is not involved in the axioms defining a vector space. One may thus replace the field of scalars by a ring R, and this gives a structure called module over R, or R-module.

The concepts of linear independence, span, basis, and linear maps (also called module homomorphisms) are defined for modules exactly as for vector spaces, with the essential difference that, if R is not a field, there are modules that do not have any basis. The modules that have a basis are the free modules, and those that are spanned by a finite set are the finitely generated modules. Module homomorphisms between finitely generated free modules may be represented by matrices. The theory of matrices over a ring is similar to that of matrices over a field, except that determinants exist only if the ring is commutative, and that a square matrix over a commutative ring is invertible only if its determinant has a multiplicative inverse in the ring.

Vector spaces are completely characterized by their dimension (up to an isomorphism). In general, there is not such a complete classification for modules, even if one restricts oneself to finitely generated modules. However, every module is a cokernel of a homomorphism of free modules.

Modules over the integers can be identified with abelian groups, since the multiplication by an integer may identified to a repeated addition. Most of the theory of abelian groups may be extended to modules over a principal ideal domain. In particular, over a principal ideal domain, every submodule of a free module is free, and the fundamental theorem of finitely generated abelian groups may be extended straightforwardly to finitely generated modules over a principal ring.

There are many rings for which there are algorithms for solving linear equations and systems of linear equations. However, these algorithms have generally a computational complexity that is much higher than the similar algorithms over a field.

Multilinear Algebra and Tensors

In multilinear algebra, one considers multivariable linear transformations, that is, mappings that are linear in each of a number of different variables. This line of inquiry naturally leads to the idea of the dual space, the vector space $V*$ consisting of linear maps $f: V \to F$ where F is the field of scalars. Multilinear maps $T: V^n \to F$ can be described via tensor products of elements of $V*$.

If, in addition to vector addition and scalar multiplication, there is a bilinear vector product $V \times V \to V$, the vector space is called an algebra; for instance, associative algebras are algebras with an associate vector product (like the algebra of square matrices, or the algebra of polynomials).

Topological Vector Spaces

Vector spaces that are not finite dimensional often require additional structure to be tractable. A normed vector space is a vector space along with a function called a norm, which measures the "size" of elements. The norm induces a metric, which measures the distance between elements, and induces a topology, which allows for a definition of continuous maps. The metric also allows for a definition of limits and completeness - a metric space that is complete is known as a Banach space. A complete metric space along with the additional structure of an inner product (a conjugate symmetric sesquilinear form) is known as a Hilbert space, which is in some sense a particularly well-behaved Banach space. Functional analysis applies the methods of linear algebra alongside those of mathematical analysis to study various function spaces; the central objects of study in functional analysis are L^p spaces, which are Banach spaces, and especially the L^2 space of square integrable functions, which is the only Hilbert space among them. Functional analysis is of particular importance to quantum mechanics, the theory of partial differential equations, digital signal processing, and electrical engineering. It also provides the foundation and theoretical framework that underlies the Fourier transform and related method.

UNIVERSAL ALGEBRA

Universal algebra studies common properties of all algebraic structures, including groups, rings, fields, lattices, etc.

A universal algebra is a pair $\mathbb{A} = \left(\mathbb{A}, \left(f_i^{\mathbb{A}} \right)_{i \in I} \right)$, where A and I are sets and for each $i \in I$, f_i^A is an operation on A. The algebra $\mathbb{A}$ is finitary if each of its operations is finitary.

A set of function symbols (or operations) of degree $n \geq 0$ is called a signature (or type). Let Σ be a signature. An algebra A is defined by a domain S (which is called its carrier or universe) and a mapping that relates a function $f: S^n \to S$ to each n-place function symbol from Σ.

Let $\mathcal{A}$ and $\mathcal{B}$ be two algebras over the same signature Σ, and their carriers are A and B, respectively. A mapping $\phi: A \to B$ is called a homomorphism from $\mathcal{A}$ to $\mathcal{B}$ if for every $f \in \Sigma$ and all $x_1, ..., x_n \in A$,

$$\phi\left(f\left(x_1, ..., x_n \right) \right) = f\left(\phi\left(x_1 \right), ..., \phi\left(x_n \right) \right).$$

If a homomorphism ϕ is surjective, then it is called epimorphism. If ϕ is an epimorphism, then B is called a homomorphic image of A. If the homomorphism ϕ is a bijection, then it is called an isomorphism. On the class of all algebras, define a relation $\sim$ by $\mathbb{A} \sim \mathbb{B}$ if and only if there is an isomorphism from $\mathbb{A}$ onto $\mathbb{B}$. Then the relation $\sim$ is an equivalence relation. Its equivalence classes are called isomorphism classes, and are typically proper classes.

A homomorphism from $\mathcal{A}$ to $\mathcal{B}$ is often denoted as $\phi\colon A \to B$. A homomorphism $\phi\colon A \to A$ is called an endomorphism. An isomorphism $\phi\colon A \to A$ is called an automorphism. The notions of homomorphism, isomorphism, endomorphism, etc., are generalizations of the respective notions in groups, rings, and other algebraic theories.

Identities (or equalities) in algebra $\mathcal{A}$ over signature Σ have the form

$$s = t,$$

where s and t are terms built up from variables using function symbols from Σ.

An identity $s = t$, is said to hold in an algebra $\mathcal{A}$ if it is true for all possible values of variables in the identity, i.e., for all possible ways of replacing the variables by elements of the carrier. The algebra $\mathcal{A}$ is then said to satisfy the identity $s = t$.

Universal algebra includes algebras such as Boolean algebras, unary algebras, etc.

Boolean Algebra

In mathematics and mathematical logic, Boolean algebra is the branch of algebra in which the values of the variables are the truth values *true* and *false*, usually denoted 1 and 0 respectively. Instead of elementary algebra where the values of the variables are numbers, and the prime operations are addition and multiplication, the main operations of Boolean algebra are the conjunction *and* denoted as $\wedge$, the disjunction *or* denoted as $\vee$, and the negation *not* denoted as $\neg$. It is thus a formalism for describing logical operations in the same way that elementary algebra describes numerical operations.

Boolean algebra has been fundamental in the development of digital electronics, and is provided for in all modern programming languages. It is also used in set theory and statistics.

Values

Whereas in elementary algebra expressions denote mainly numbers, in Boolean algebra they denote the truth values *false* and *true*. These values are represented with the bits (or binary digits), namely 0 and 1. They do not behave like the integers 0 and 1, for which $1 + 1 = 2$, but may be identified with the elements of the two-element field GF(2), that is, integer arithmetic modulo 2, for which $1 + 1 = 0$. Addition and multiplication then play the Boolean roles of XOR (exclusive-or) and AND (conjunction) respectively, with disjunction $x \vee y$ (inclusive-or) definable as $x + y - xy$.

Boolean algebra also deals with functions which have their values in the set $\{0, 1\}$. A sequence of bits is a commonly used such function. Another common example is the subsets of a set E: to a subset F of E is associated the indicator function that takes the value 1 on F and 0 outside F. The most general example is the elements of a Boolean algebra, with all of the foregoing being instances thereof.

As with elementary algebra, the purely equational part of the theory may be developed without considering explicit values for the variables.

Operations

Basic operations

The basic operations of Boolean algebra are as follows:

- AND (conjunction), denoted $x \wedge y$ (sometimes x AND y or Kxy), satisfies $x \wedge y = 1$ if $x = y = 1$, and $x \wedge y = 0$ otherwise.

- OR (disjunction), denoted $x \vee y$ (sometimes x OR y or Axy), satisfies $x \vee y = 0$ if $x = y = 0$, and $x \vee y = 1$ otherwise.

- NOT (negation), denoted $\neg x$ (sometimes NOT x, Nx or $!x$), satisfies $\neg x = 0$ if $x = 1$ and $\neg x = 1$ if $x = 0$.

Alternatively the values of $x \wedge y$, $x \vee y$, and $\neg x$ can be expressed by tabulating their values with truth tables as follows:

x	y	$x \wedge y$	$x \vee y$	x	$\neg x$
0	0	0	0	0	1
1	0	0	1	1	0
0	1	0	1		
1	1	1	1		

If the truth values 0 and 1 are interpreted as integers, these operations may be expressed with the ordinary operations of arithmetic (where $x + y$ uses addition and xy uses multiplication), or by the minimum/maximum functions:

$$x \wedge y = xy = \min(x, y)$$
$$x \vee y = x + y - xy = \max(x, y)$$
$$\neg x = 1 - x$$

One might consider that only negation and one of the two other operations are basic, because of the following identities that allow one to define conjunction in terms of negation and the disjunction, and vice versa (De Morgan's laws):

$$x \wedge y = \neg(\neg x \vee \neg y)$$
$$x \vee y = \neg(\neg x \wedge \neg y)$$

Secondary Operations

The three Boolean operations described above are referred to as basic, meaning that they can be taken as a basis for other Boolean operations that can be built up from them by composition, the manner in which operations are combined or compounded. Operations composed from the basic operations include the following examples:

$$x \rightarrow y = \neg x \vee y$$
$$x \oplus y = (x \vee y) \wedge \neg(x \wedge y) = (x \wedge \neg y) \vee (\neg x \wedge y)$$
$$x \equiv y = \neg(x \oplus y) = (x \wedge y) \vee (\neg x \wedge \neg y)$$

These definitions give rise to the following truth tables giving the values of these operations for all four possible inputs.

Secondary operations				
x	y	$x \rightarrow y$	$x \oplus y$	$x \equiv y$
0	0	1	0	1
1	0	0	1	0
0	1	1	1	0
1	1	1	0	1

The first operation, $x \rightarrow y$, or Cxy, is called material implication. If x is true then the value of $x \rightarrow y$ is taken to be that of y (e.g. if x is true and y is false, then $x \rightarrow y$ is also false). But if x is false then the value of y can be ignored; however the operation must return *some* boolean value and there are only two choices. So by definition, $x \rightarrow y$ is *true* when x is false. (Relevance logic suggests this definition by viewing an implication with a false premise as something other than either true or false).

The second operation, $x \oplus y$, or Jxy, is called exclusive or (often abbreviated as XOR) to distinguish it from disjunction as the inclusive kind. It excludes the possibility of both x and y *being* true: if both are true then result is false. Defined in terms of arithmetic it is addition mod 2 where $1 + 1 = 0$.

The third operation, the complement of exclusive or, is equivalence or Boolean equality: $x \equiv y$, or Exy, is true just when x and y have the same value. Hence $x \oplus y$ as its complement can be understood as $x \neq y$, being true just when x and y are different. Equivalence's counterpart in arithmetic mod 2 is $x + y + 1$.

Given two operands, each with two possible values, there are $2^2 = 4$ possible combinations of inputs. Because each output can have two possible values, there are a total of $2^4 = 16$ possible binary Boolean operations. Any such operation or function (as well as any Boolean function with more inputs) can be expressed with the basic operations from above. Hence the basic operations are functionally complete.

Laws

A law of Boolean algebra is an identity such as $x \vee (y \vee z) = (x \vee y) \vee z$ between two Boolean terms, where a Boolean term is defined as an expression built up from variables and the constants 0 and 1 using the operations $\wedge$, $\vee$, and $\neg$. The concept can be extended to terms involving other Boolean operations such as $\oplus$, $\rightarrow$, and $\equiv$, but such extensions are unnecessary for the purposes to which the laws are put. Such purposes include the definition of a Boolean algebra as any model of the Boolean laws, and as a means for deriving new laws from old as in the derivation of $x \vee (y \wedge z) = x \vee (z \wedge y)$ from $y \wedge z = z \wedge y$ as treated in the section on axiomatization.

Monotone Laws

Boolean algebra satisfies many of the same laws as ordinary algebra when one matches up $\vee$ with addition and ∧ with multiplication. In particular the following laws are common to both kinds of algebra:

Associativity of ∨ :	$x \vee (y \vee z)$	$= (x \vee y) \vee z$
Associativity of ∧ :	$x \wedge (y \wedge z)$	$= (x \wedge y) \wedge z$
Commutativity of ∨ :	$= y \vee x$	$= y \vee x$
Commutativity of ∧ :	$x \wedge y$	$= y \wedge x$
Distributivity of ∧ over ∨ :	$x \wedge (y \vee z)$	$= (x \wedge y) \vee (x \wedge z)$
Identity for ∨ :	$x \vee 0$	$= x$
Identity for ∧ :	$x \wedge 1$	$= x$
Annihilator for ∧ :	$x \wedge 0$	$= 0$

The following laws hold in Boolean Algebra, but not in ordinary algebra:

Annihilator for ∨ :	$x \vee 1$	$= 1$
Idempotence of ∨ :	$x \vee x$	$= x$
Idempotence of ∧ :	$x \wedge x$	$= x$
Absorption 1:	$x \wedge (x \vee y)$	$= x$
Absorption 2:	$x \vee (x \wedge y)$	$= x$
Distributivity of ∨ over ∧ :	$x \vee (y \wedge z)$	$= (x \vee y) \wedge (x \vee z)$

Taking x = 2 in the third law above shows that it is not an ordinary algebra law, since 2×2 = 4. The remaining five laws can be falsified in ordinary algebra by taking all variables to be 1, for example in Absorption Law 1 the left hand side would be 1(1+1) = 2 while the right hand side would be 1, and so on.

All of the laws treated so far have been for conjunction and disjunction. These operations have the property that changing either argument either leaves the output unchanged or the output changes in the same way as the input. Equivalently, changing any variable from 0 to 1 never results in the output changing from 1 to 0. Operations with this property are said to be monotone. Thus the axioms so far have all been for monotonic Boolean logic. Nonmonotonicity enters via complement ¬ as follows.

Nonmonotone Laws

The complement operation is defined by the following two laws:

Complementation 1 $x \wedge \neg x = 0$
Complementation 2 $x \vee \neg x = 1$

All properties of negation including the laws below follow from the above two laws alone.

In both ordinary and Boolean algebra, negation works by exchanging pairs of elements, whence in both algebras it satisfies the double negation law (also called involution law).

Double negation $\neg(\neg x) = x$

But whereas *ordinary algebra* satisfies the two laws:

$$(-x)(-y) = xy$$
$$(-x) + (-y) = -(x + y)$$

Boolean algebra satisfies De Morgan's laws:

De Morgan 1 $\neg x \wedge \neg y = \neg(x \vee y)$

De Morgan 2 $\neg x \vee \neg y = \neg(x \wedge y)$

Completeness

The laws listed above define Boolean algebra, in the sense that they entail the rest of the subject. The laws *Complementation* 1 and 2, together with the monotone laws, suffice for this purpose and can therefore be taken as one possible *complete* set of laws or axiomatization of Boolean algebra. Every law of Boolean algebra follows logically from these axioms. Furthermore, Boolean algebras can then be defined as the models of these axioms as treated in the section thereon.

To clarify, writing down further laws of Boolean algebra cannot give rise to any new consequences of these axioms, nor can it rule out any model of them. In contrast, in a list of some but not all of the same laws, there could have been Boolean laws that did not follow from those on the list, and moreover there would have been models of the listed laws that were not Boolean algebras.

This axiomatization is by no means the only one, or even necessarily the most natural given that we did not pay attention to whether some of the axioms followed from others but simply chose to stop when we noticed we had enough laws, treated further in the section on axiomatizations. Or the intermediate notion of axiom can be sidestepped altogether by defining a Boolean law directly as any tautology, understood as an equation that holds for all values of its variables over 0 and 1. All these definitions of Boolean algebra can be shown to be equivalent.

Duality Principle

Principle: If {X, R} is a poset, then {X, R(inverse)} is also a poset.

There is nothing magical about the choice of symbols for the values of Boolean algebra. We could rename 0 and 1 to say α and β, and as long as we did so consistently throughout it would still be Boolean algebra, albeit with some obvious cosmetic differences.

But suppose we rename 0 and 1 to 1 and 0 respectively. Then it would still be Boolean algebra, and moreover operating on the same values. However it would not be identical to our original Boolean algebra because now we find $\vee$ behaving the way $\wedge$ used to do and vice versa. So there are still some cosmetic differences to show that we've been fiddling with the notation, despite the fact that we're still using 0s and 1s.

But if in addition to interchanging the names of the values we also interchange the names of the two binary operations, *now* there is no trace of what we have done. The end product is completely indistinguishable from what we started with. We might notice that the columns for $x \wedge y$ and $x \vee y$ in the truth tables had changed places, but that switch is immaterial.

When values and operations can be paired up in a way that leaves everything important unchanged when all pairs are switched simultaneously, we call the members of each pair dual to each other. Thus 0 and 1 are dual, and $\wedge$ and $\vee$ are dual. The Duality Principle, also called De Morgan duality, asserts that Boolean algebra is unchanged when all dual pairs are interchanged.

One change we did not need to make as part of this interchange was to complement. We say that complement is a self-dual operation. The identity or do-nothing operation x (copy the input to the output) is also self-dual. A more complicated example of a self-dual operation is $(x \wedge y) \vee (y \wedge z) \vee (z \wedge x)$. There is no self-dual binary operation that depends on both its arguments. A composition of self-dual operations is a self-dual operation. For example, if $f(x, y, z) = (x \wedge y) \vee (y \wedge z) \vee (z \wedge x)$, then $f(f(x, y, z), x, t)$ is a self-dual operation of four arguments x,y,z,t.

The principle of duality can be explained from a group theory perspective by the fact that there are exactly four functions that are one-to-one mappings (automorphisms) of the set of Boolean polynomials back to itself: the identity function, the complement function, the dual function and the contradual function (complemented dual). These four functions form a group under function composition, isomorphic to the Klein four-group, acting on the set of Boolean polynomials. Walter Gottschalk remarked that consequently a more appropriate name for the phenomenon would be the *principle* (or *square*) *of quaternality*.

Diagrammatic Representations

Venn Diagrams

A Venn diagram is a representation of a Boolean operation using shaded overlapping regions. There is one region for each variable, all circular in the examples here. The interior and exterior of region x corresponds respectively to the values 1 (true) and 0 (false) for variable x. The shading indicates the value of the operation for each combination of regions, with dark denoting 1 and light 0 (some authors use the opposite convention).

The three Venn diagrams in the figure below represent respectively conjunction $x \wedge y$, disjunction $x \vee y$, and complement $\neg x$.

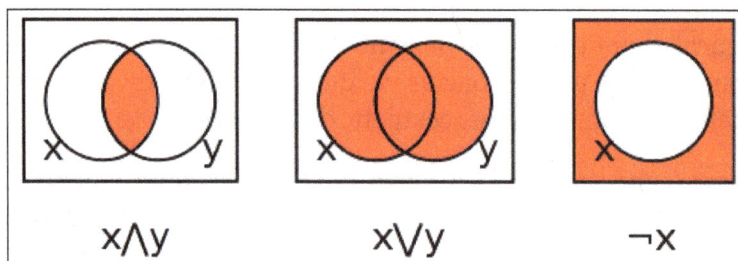

Venn diagrams for conjunction, disjunction, and complement.

For conjunction, the region inside both circles is shaded to indicate that $x \wedge y$ is 1 when both variables are 1. The other regions are left unshaded to indicate that $x \wedge y$ is 0 for the other three combinations.

The second diagram represents disjunction $x \lor y$ by shading those regions that lie inside either or both circles. The third diagram represents complement $\neg x$ by shading the region *not* inside the circle.

While we have not shown the Venn diagrams for the constants 0 and 1, they are trivial, being respectively a white box and a dark box, neither one containing a circle. However we could put a circle for x in those boxes, in which case each would denote a function of one argument, x, which returns the same value independently of x, called a constant function. As far as their outputs are concerned, constants and constant functions are indistinguishable; the difference is that a constant takes no arguments, called a *zeroary* or *nullary* operation, while a constant function takes one argument, which it ignores, and is a *unary* operation.

Venn diagrams are helpful in visualizing laws. The commutativity laws for $\land$ and $\lor$ can be seen from the symmetry of the diagrams: a binary operation that was not commutative would not have a symmetric diagram because interchanging x and y would have the effect of reflecting the diagram horizontally and any failure of commutativity would then appear as a failure of symmetry.

Idempotence of $\land$ and $\lor$ can be visualized by sliding the two circles together and noting that the shaded area then becomes the whole circle, for both $\land$ and $\lor$.

To see the first absorption law, $x \land (x \lor y) = x$, start with the diagram in the middle for $x \lor y$ and note that the portion of the shaded area in common with the x circle is the whole of the x circle. For the second absorption law, $x \lor (x \land y) = x$, start with the left diagram for $x \land y$ and note that shading the whole of the x circle results in just the x circle being shaded, since the previous shading was inside the x circle.

The double negation law can be seen by complementing the shading in the third diagram for $\neg x$, which shades the x circle.

To visualize the first De Morgan's law, $(\neg x) \land (\neg y) = \neg(x \lor y)$, start with the middle diagram for $x \lor y$ and complement its shading so that only the region outside both circles is shaded, which is what the right hand side of the law describes. The result is the same as if we shaded that region which is both outside the x circle *and* outside the y circle, i.e. the conjunction of their exteriors, which is what the left hand side of the law describes.

The second De Morgan's law, $(\neg x) \lor (\neg y) = \neg(x \land y)$, works the same way with the two diagrams interchanged.

The first complement law, $x \land \neg x = 0$, says that the interior and exterior of the x circle have no overlap. The second complement law, $x \lor \neg x = 1$, says that everything is either inside or outside the x circle.

Digital Logic Gates

Digital logic is the application of the Boolean algebra of 0 and 1 to electronic hardware consisting of logic gates connected to form a circuit diagram. Each gate implements a Boolean operation, and is depicted schematically by a shape indicating the operation. The shapes associated with the gates for conjunction (AND-gates), disjunction (OR-gates), and complement (inverters) are as follows.

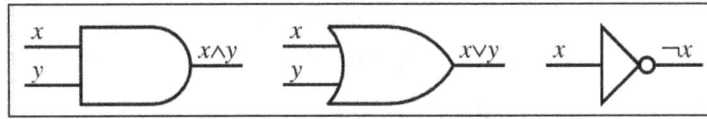

From left to right: AND, OR, and NOT gates.

The lines on the left of each gate represent input wires or *ports*. The value of the input is represented by a voltage on the lead. For so-called "active-high" logic, 0 is represented by a voltage close to zero or "ground", while 1 is represented by a voltage close to the supply voltage; active-low reverses this. The line on the right of each gate represents the output port, which normally follows the same voltage conventions as the input ports.

Complement is implemented with an inverter gate. The triangle denotes the operation that simply copies the input to the output; the small circle on the output denotes the actual inversion complementing the input. The convention of putting such a circle on any port means that the signal passing through this port is complemented on the way through, whether it is an input or output port.

The Duality Principle, or De Morgan's laws, can be understood as asserting that complementing all three ports of an AND gate converts it to an OR gate and vice versa, as shown in Figure below. Complementing both ports of an inverter however leaves the operation unchanged.

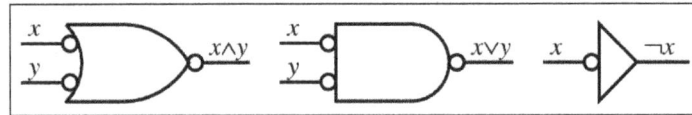

More generally one may complement any of the eight subsets of the three ports of either an AND or OR gate. The resulting sixteen possibilities give rise to only eight Boolean operations, namely those with an odd number of 1's in their truth table. There are eight such because the "odd-bit-out" can be either 0 or 1 and can go in any of four positions in the truth table. There being sixteen binary Boolean operations, this must leave eight operations with an even number of 1's in their truth tables. Two of these are the constants 0 and 1 (as binary operations that ignore both their inputs); four are the operations that depend nontrivially on exactly one of their two inputs, namely x, y, $\neg x$, and $\neg y$; and the remaining two are $x \oplus y$ (XOR) and its complement $x \equiv y$.

Boolean Algebras

The term "algebra" denotes both a subject, namely the subject of algebra, and an object, namely an algebraic structure.

Concrete Boolean Algebras

A concrete Boolean algebra or field of sets is any nonempty set of subsets of a given set X closed under the set operations of union, intersection, and complement relative to X.

(As an aside, historically X itself was required to be nonempty as well to exclude the degenerate or one-element Boolean algebra, which is the one exception to the rule that all Boolean algebras satisfy the same equations since the degenerate algebra satisfies every equation. However this exclusion conflicts with the preferred purely equational definition of "Boolean algebra," there being no way to rule out the one-element algebra using only equations— 0 ≠ 1 does not count, being a negated equation. Hence modern authors allow the degenerate Boolean algebra and let X be empty).

Example: The power set 2^X of X, consisting of all subsets of X. Here X may be any set: empty, finite, infinite, or even uncountable.

Example: The empty set and X. This two-element algebra shows that a concrete Boolean algebra can be finite even when it consists of subsets of an infinite set. It can be seen that every field of subsets of X must contain the empty set and X. Hence no smaller example is possible, other than the degenerate algebra obtained by taking X to be empty so as to make the empty set and X coincide.

Example: The set of finite and cofinite sets of integers, where a cofinite set is one omitting only finitely many integers. This is clearly closed under complement, and is closed under union because the union of a cofinite set with any set is cofinite, while the union of two finite sets is finite. Intersection behaves like union with "finite" and "cofinite" interchanged.

Example: For a less trivial example of the point made by Example 2, consider a Venn diagram formed by n closed curves partitioning the diagram into 2^n regions, and let X be the (infinite) set of all points in the plane not on any curve but somewhere within the diagram. The interior of each region is thus an infinite subset of X, and every point in X is in exactly one region. Then the set of all 2^{2^n} possible unions of regions (including the empty set obtained as the union of the empty set of regions and X obtained as the union of all 2^n regions) is closed under union, intersection, and complement relative to X and therefore forms a concrete Boolean algebra. Again we have finitely many subsets of an infinite set forming a concrete Boolean algebra, with Example 2 arising as the case $n = 0$ of no curves.

Subsets as Bit Vectors

A subset Y of X can be identified with an indexed family of bits with index set X, with the bit indexed by $x \in X$ being 1 or 0 according to whether or not $x \in Y$. (This is the so-called characteristic function notion of a subset.) For example, a 32-bit computer word consists of 32 bits indexed by the set {0,1,2,...,31}, with 0 and 31 indexing the low and high order bits respectively. For a smaller example, if $X = \{a,b,c\}$ where a, b, c are viewed as bit positions in that order from left to right, the eight subsets {}, {c}, {b}, {b,c}, {a}, {a,c}, {a,b}, and {a,b,c} of X can be identified with the respective bit vectors 000, 001, 010, 011, 100, 101, 110, and 111. Bit vectors indexed by the set of natural numbers are infinite sequences of bits, while those indexed by the reals in the unit interval [0,1] are packed too densely to be able to write conventionally but nonetheless form well-defined indexed families (imagine coloring every point of the interval [0,1] either black or white independently; the black points then form an arbitrary subset of [0,1]).

From this bit vector viewpoint, a concrete Boolean algebra can be defined equivalently as a non-empty set of bit vectors all of the same length (more generally, indexed by the same set) and closed under the bit vector operations of bitwise ∧, ∨, and ¬, as in 1010∧0110 = 0010, 1010∨0110 = 1110, and ¬1010 = 0101, the bit vector realizations of intersection, union, and complement respectively.

The Prototypical Boolean Algebra

The set {0,1} and its Boolean operations as treated above can be understood as the special case of bit vectors of length one, which by the identification of bit vectors with subsets can also be understood as the two subsets of a one-element set. We call this the prototypical Boolean algebra, justified by the following observation.

> The laws satisfied by all nondegenerate concrete Boolean algebras coincide with those satisfied by the prototypical Boolean algebra.

This observation is easily proved as follows. Certainly any law satisfied by all concrete Boolean algebras is satisfied by the prototypical one since it is concrete. Conversely any law that fails for some concrete Boolean algebra must have failed at a particular bit position, in which case that position by itself furnishes a one-bit counterexample to that law. Nondegeneracy ensures the existence of at least one bit position because there is only one empty bit vector.

Boolean Algebras

The Boolean algebras we have seen so far have all been concrete, consisting of bit vectors or equivalently of subsets of some set. Such a Boolean algebra consists of a set and operations on that set which can be *shown* to satisfy the laws of Boolean algebra.

Instead of showing that the Boolean laws are satisfied, we can instead postulate a set X, two binary operations on X, and one unary operation, and *require* that those operations satisfy the laws of Boolean algebra. The elements of X need not be bit vectors or subsets but can be anything at all. This leads to the more general *abstract* definition.

> A Boolean algebra is any set with binary operations $\wedge$ and $\vee$ and a unary operation $\neg$ thereon satisfying the Boolean laws.

For the purposes of this definition it is irrelevant how the operations came to satisfy the laws, whether by fiat or proof. All concrete Boolean algebras satisfy the laws (by proof rather than fiat), whence every concrete Boolean algebra is a Boolean algebra according to our definitions. This axiomatic definition of a Boolean algebra as a set and certain operations satisfying certain laws or axioms *by fiat* is entirely analogous to the abstract definitions of group, ring, field etc. characteristic of modern or abstract algebra.

Given any complete axiomatization of Boolean algebra, such as the axioms for a complemented distributive lattice, a sufficient condition for an algebraic structure of this kind to satisfy all the Boolean laws is that it satisfy just those axioms. The following is therefore an equivalent definition:

> A Boolean algebra is a complemented distributive lattice.

Representable Boolean Algebras

Although every concrete Boolean algebra is a Boolean algebra, not every Boolean algebra need be concrete. Let n be a square-free positive integer, one not divisible by the square of an integer, for example 30 but not 12. The operations of greatest common divisor, least common multiple, and division into n (that is, $\neg x = n/x$), can be shown to satisfy all the Boolean laws when their arguments range over the positive divisors of n. Hence those divisors form a Boolean algebra. These divisors are not subsets of a set, making the divisors of n a Boolean algebra that is not concrete according to our definitions.

However, if we *represent* each divisor of n by the set of its prime factors, we find that this nonconcrete Boolean algebra is isomorphic to the concrete Boolean algebra consisting of all sets of prime factors of n, with union corresponding to least common multiple, intersection to greatest common divisor, and complement to division into n. So this example while not technically concrete is at least "morally" concrete via this representation, called an isomorphism. This example is an instance of the following notion:

> A Boolean algebra is called representable when it is isomorphic to a concrete Boolean algebra.

The obvious next question is answered positively as follows:

> Every Boolean algebra is representable.

That is, up to isomorphism, abstract and concrete Boolean algebras are the same thing. This quite nontrivial result depends on the Boolean prime ideal theorem, a choice principle slightly weaker than the axiom of choice. This strong relationship implies a weaker result strengthening the observation in the previous subsection to the following easy consequence of representability.

> The laws satisfied by all Boolean algebras coincide with those satisfied by the prototypical Boolean algebra.

It is weaker in the sense that it does not of itself imply representability. Boolean algebras are special here, for example a relation algebra is a Boolean algebra with additional structure but it is not the case that every relation algebra is representable in the sense appropriate to relation algebras.

Axiomatizing Boolean Algebra

The above definition of an abstract Boolean algebra as a set and operations satisfying "the" Boolean laws raises the question, what are those laws? A simple-minded answer is "all Boolean laws," which can be defined as all equations that hold for the Boolean algebra of 0 and 1. Since there are infinitely many such laws this is not a terribly satisfactory answer in practice, leading to the next question: does it suffice to require only finitely many laws to hold?

In the case of Boolean algebras the answer is yes. In particular the finitely many equations we have listed above suffice. We say that Boolean algebra is finitely axiomatizable or finitely based.

Can this list be made shorter yet? Again the answer is yes. To begin with, some of the above laws are implied by some of the others. A sufficient subset of the above laws consists of the pairs of associativity, commutativity, and absorption laws, distributivity of $\land$ over $\lor$ (or the other distributivity law—one suffices), and the two complement laws. In fact this is the traditional axiomatization of Boolean algebra as a complemented distributive lattice.

By introducing additional laws not listed above it becomes possible to shorten the list yet further. In 1933, Edward Huntington showed that if the basic operations are taken to be $x \lor y$ and $\neg x$, with $x \land y$ considered a derived operation (e.g. via De Morgan's law in the form $x \land y = \neg(\neg x \lor \neg y)$), then the equation $\neg(\neg x \lor \neg y) \lor \neg(\neg x \lor y) = x$ along with the two equations expressing associativity and commutativity of $\lor$ completely axiomatized Boolean algebra. When the only basic operation is the binary NAND operation $\neg(x \land y)$, Stephen Wolfram has proposed the single axiom $((xy)z)(x((xz)x))$

= z as a one-equation axiomatization of Boolean algebra, where for convenience here xy denotes the NAND rather than the AND of x and y.

Propositional Logic

Propositional logic is a logical system that is intimately connected to Boolean algebra. Many syntactic concepts of Boolean algebra carry over to propositional logic with only minor changes in notation and terminology, while the semantics of propositional logic are defined via Boolean algebras in a way that the tautologies (theorems) of propositional logic correspond to equational theorems of Boolean algebra.

Syntactically, every Boolean term corresponds to a propositional formula of propositional logic. In this translation between Boolean algebra and propositional logic, Boolean variables x, y... become propositional variables (or atoms) $P,Q,...$, Boolean terms such as $x \vee y$ become propositional formulas $P \vee Q$, 0 becomes *false* or $\perp$, and 1 becomes *true* or T. It is convenient when referring to generic propositions to use Greek letters Φ, Ψ,... as metavariables (variables outside the language of propositional calculus, used when talking *about* propositional calculus) to denote propositions.

The semantics of propositional logic rely on truth assignments. The essential idea of a truth assignment is that the propositional variables are mapped to elements of a fixed Boolean algebra, and then the truth value of a propositional formula using these letters is the element of the Boolean algebra that is obtained by computing the value of the Boolean term corresponding to the formula. In classical semantics, only the two-element Boolean algebra is used, while in Boolean-valued semantics arbitrary Boolean algebras are considered. A tautology is a propositional formula that is assigned truth value *1* by every truth assignment of its propositional variables to an arbitrary Boolean algebra (or, equivalently, every truth assignment to the two element Boolean algebra).

These semantics permit a translation between tautologies of propositional logic and equational theorems of Boolean algebra. Every tautology Φ of propositional logic can be expressed as the Boolean equation $\Phi = 1$, which will be a theorem of Boolean algebra. Conversely every theorem $\Phi = \Psi$ of Boolean algebra corresponds to the tautologies $(\Phi \vee \neg \Psi) \wedge (\neg \Phi \vee \Psi)$ and $(\Phi \wedge \Psi) \vee (\neg \Phi \wedge \neg \Psi)$. If $\rightarrow$ is in the language these last tautologies can also be written as $(\Phi \rightarrow \Psi) \wedge (\Psi \rightarrow \Phi)$, or as two separate theorems $\Phi \rightarrow \Psi$ and $\Psi \rightarrow \Phi$; if $\equiv$ is available then the single tautology $\Phi \equiv \Psi$ can be used.

Applications

One motivating application of propositional calculus is the analysis of propositions and deductive arguments in natural language. Whereas the proposition "if $x = 3$ then $x+1 = 4$" depends on the meanings of such symbols as + and 1, the proposition "if $x = 3$ then $x = 3$" does not; it is true merely by virtue of its structure, and remains true whether "$x = 3$" is replaced by "$x = 4$" or "the moon is made of green cheese." The generic or abstract form of this tautology is "if P then P", or in the language of Boolean algebra, "$P \rightarrow P$".

Replacing P by $x = 3$ or any other proposition is called instantiation of P by that proposition. The result of instantiating P in an abstract proposition is called an instance of the proposition. Thus "$x = 3 \rightarrow x = 3$" is a tautology by virtue of being an instance of the abstract tautology "$P \rightarrow P$". All occurrences of the instantiated variable must be instantiated with the same proposition, to avoid such nonsense as $P \to x = 3$ or $x = 3 \to x = 4$.

Propositional calculus restricts attention to abstract propositions, those built up from propositional variables using Boolean operations. Instantiation is still possible within propositional calculus, but only by instantiating propositional variables by abstract propositions, such as instantiating Q by $Q \to P$ in $P \to (Q \to P)$ to yield the instance $P \to ((Q \to P) \to P)$.

(The availability of instantiation as part of the machinery of propositional calculus avoids the need for metavariables within the language of propositional calculus, since ordinary propositional variables can be considered within the language to denote arbitrary propositions. The metavariables themselves are outside the reach of instantiation, not being part of the language of propositional calculus but rather part of the same language for talking about it that this sentence is written in, where we need to be able to distinguish propositional variables and their instantiations as being distinct syntactic entities).

Deductive Systems for Propositional Logic

An axiomatization of propositional calculus is a set of tautologies called axioms and one or more inference rules for producing new tautologies from old. A *proof* in an axiom system A is a finite nonempty sequence of propositions each of which is either an instance of an axiom of A or follows by some rule of A from propositions appearing earlier in the proof (thereby disallowing circular reasoning). The last proposition is the theorem proved by the proof. Every nonempty initial segment of a proof is itself a proof, whence every proposition in a proof is itself a theorem. An axiomatization is sound when every theorem is a tautology, and complete when every tautology is a theorem.

Sequent Calculus

Propositional calculus is commonly organized as a Hilbert system, whose operations are just those of Boolean algebra and whose theorems are Boolean tautologies, those Boolean terms equal to the Boolean constant 1. Another form is sequent calculus, which has two sorts, propositions as in ordinary propositional calculus, and pairs of lists of propositions called sequents, such as $A \lor B, A \land C, \ldots \vdash A, B \to C, \ldots$. The two halves of a sequent are called the antecedent and the succedent respectively. The customary metavariable denoting an antecedent or part thereof is Γ, and for a succedent Δ; thus $\Gamma, A \vdash \Delta$ would denote a sequent whose succedent is a list Δ and whose antecedent is a list Γ with an additional proposition A appended after it. The antecedent is interpreted as the conjunction of its propositions, the succedent as the disjunction of its propositions, and the sequent itself as the entailment of the succedent by the antecedent.

Entailment differs from implication in that whereas the latter is a binary *operation* that returns a value in a Boolean algebra, the former is a binary *relation* which either holds or does not hold. In this sense entailment is an *external* form of implication, meaning external to the Boolean algebra, thinking of the reader of the sequent as also being external and interpreting and comparing antecedents and succedents in some Boolean algebra. The natural interpretation of $\vdash$ is as $\leq$ in the partial order of the Boolean algebra defined by $x \leq y$ just when $x \lor y = y$. This ability to mix external implication $\to$ and internal implication $\to$ in the one logic is among the essential differences between sequent calculus and propositional calculus.

Applications

Boolean algebra as the calculus of two values is fundamental to computer circuits, computer programming, and mathematical logic, and is also used in other areas of mathematics such as set theory and statistics.

Computers

In the early 20th century, several electrical engineers intuitively recognized that Boolean algebra was analogous to the behavior of certain types of electrical circuits. Claude Shannon formally proved such behavior was logically equivalent to Boolean algebra in his 1937 master's thesis, *A Symbolic Analysis of Relay and Switching Circuits.*

Today, all modern general purpose computers perform their functions using two-value Boolean logic; that is, their electrical circuits are a physical manifestation of two-value Boolean logic. They achieve this in various ways: as voltages on wires in high-speed circuits and capacitive storage devices, as orientations of a magnetic domain in ferromagnetic storage devices, as holes in punched cards or paper tape, and so on. (Some early computers used decimal circuits or mechanisms instead of two-valued logic circuits).

Of course, it is possible to code more than two symbols in any given medium. For example, one might use respectively 0, 1, 2, and 3 volts to code a four-symbol alphabet on a wire, or holes of different sizes in a punched card. In practice, the tight constraints of high speed, small size, and low power combine to make noise a major factor. This makes it hard to distinguish between symbols when there are several possible symbols that could occur at a single site. Rather than attempting to distinguish between four voltages on one wire, digital designers have settled on two voltages per wire, high and low.

Computers use two-value Boolean circuits for the above reasons. The most common computer architectures use ordered sequences of Boolean values, called bits, of 32 or 64 values, e.g. 011010001 1010110010101010101001011. When programming in machine code, assembly language, and certain other programming languages, programmers work with the low-level digital structure of the data registers. These registers operate on voltages, where zero volts represents Boolean 0, and a reference voltage (often +5V, +3.3V, +1.8V) represents Boolean 1. Such languages support both numeric operations and logical operations. In this context, "numeric" means that the computer treats sequences of bits as binary numbers (base two numbers) and executes arithmetic operations like add, subtract, multiply, or divide. "Logical" refers to the Boolean logical operations of disjunction, conjunction, and negation between two sequences of bits, in which each bit in one sequence is simply compared to its counterpart in the other sequence. Programmers therefore have the option of working in and applying the rules of either numeric algebra or Boolean algebra as needed. A core differentiating feature between these families of operations is the existence of the carry operation in the first but not the second.

Two-valued Logic

Other areas where two values is a good choice are the law and mathematics. In everyday relaxed conversation, nuanced or complex answers such as "maybe" or "only on the weekend" are acceptable. In more focused situations such as a court of law or theorem-based mathematics however it is deemed advantageous to frame questions so as to admit a simple yes-or-no answer—is the defendant guilty or not guilty, is the proposition true or false—and to disallow any other answer. However much of a straitjacket this might prove in practice for the respondent, the principle of the simple yes-no question has become a central feature of both judicial and mathematical logic, making two-valued logic deserving of organization and study in its own right.

A central concept of set theory is membership. Now an organization may permit multiple degrees of membership, such as novice, associate, and full. With sets however an element is either in or out. The candidates for membership in a set work just like the wires in a digital computer: each candidate is either a member or a nonmember, just as each wire is either high or low.

Algebra being a fundamental tool in any area amenable to mathematical treatment, these considerations combine to make the algebra of two values of fundamental importance to computer hardware, mathematical logic, and set theory.

Two-valued logic can be extended to multi-valued logic, notably by replacing the Boolean domain {0, 1} with the unit interval [0,1], in which case rather than only taking values 0 or 1, any value between and including 0 and 1 can be assumed. Algebraically, negation (NOT) is replaced with $1 - x$, conjunction (AND) is replaced with multiplication (xy), and disjunction (OR) is defined via De Morgan's law. Interpreting these values as logical truth values yields a multi-valued logic, which forms the basis for fuzzy logic and probabilistic logic. In these interpretations, a value is interpreted as the "degree" of truth – to what extent a proposition is true, or the probability that the proposition is true.

Boolean Operations

The original application for Boolean operations was mathematical logic, where it combines the truth values, true or false, of individual formulas.

Natural languages such as English have words for several Boolean operations, in particular conjunction (*and*), disjunction (*or*), negation (*not*), and implication (*implies*). *But not* is synonymous with *and not*. When used to combine situational assertions such as "the block is on the table" and "cats drink milk," which naively are either true or false, the meanings of these logical connectives often have the meaning of their logical counterparts. However, with descriptions of behavior such as "Jim walked through the door", one starts to notice differences such as failure of commutativity, for example the conjunction of "Jim opened the door" with "Jim walked through the door" in that order is not equivalent to their conjunction in the other order, since *and* usually means *and then* in such cases. Questions can be similar: the order "Is the sky blue, and why is the sky blue?" makes more sense than the reverse order. Conjunctive commands about behavior are like behavioral assertions, as in *get dressed and go to school*. Disjunctive commands such *love me or leave me* or *fish or cut bait* tend to be asymmetric via the implication that one alternative is less preferable. Conjoined nouns such as *tea and milk* generally describe aggregation as with set union while *tea or milk* is a choice. However context can reverse these senses, as in *your choices are coffee and tea* which usually means the same as *your choices are coffee or tea* (alternatives). Double negation as in "I don't not like milk" rarely means literally "I do like milk" but rather conveys some sort of hedging, as though to imply that there is a third possibility. "Not not P" can be loosely interpreted as "surely P", and although P necessarily implies "not not P" the converse is suspect in English, much as with intuitionistic logic. In view of the highly idiosyncratic usage of conjunctions in natural languages, Boolean algebra cannot be considered a reliable framework for interpreting them.

Boolean operations are used in digital logic to combine the bits carried on individual wires, thereby interpreting them over $\{0,1\}$. When a vector of n identical binary gates are used to combine two bit vectors each of n bits, the individual bit operations can be understood collectively as a single operation on values from a Boolean algebra with 2^n elements.

Naive set theory interprets Boolean operations as acting on subsets of a given set X. As we saw earlier this behavior exactly parallels the coordinate-wise combinations of bit vectors, with the union of two sets corresponding to the disjunction of two bit vectors and so on.

The 256-element free Boolean algebra on three generators is deployed in computer displays based on raster graphics, which use bit blit to manipulate whole regions consisting of pixels, relying on Boolean operations to specify how the source region should be combined with the destination, typically with the help of a third region called the mask. Modern video cards offer all $2^{23} = 256$ ternary operations for this purpose, with the choice of operation being a one-byte (8-bit) parameter. The constants SRC = 0xaa or 10101010, DST = 0xcc or 11001100, and MSK = 0xf0 or 11110000 allow Boolean operations such as (SRC^DST)&MSK (meaning XOR the source and destination and then AND the result with the mask) to be written directly as a constant denoting a byte calculated at compile time, 0x60 in the (SRC^DST)&MSK example, 0x66 if just SRC^DST, etc. At run time the video card interprets the byte as the raster operation indicated by the original expression in a uniform way that requires remarkably little hardware and which takes time completely independent of the complexity of the expression.

Solid modeling systems for computer aided design offer a variety of methods for building objects from other objects, combination by Boolean operations being one of them. In this method the space in which objects exist is understood as a set S of voxels (the three-dimensional analogue of pixels in two-dimensional graphics) and shapes are defined as subsets of S, allowing objects to be combined as sets via union, intersection, etc. One obvious use is in building a complex shape from simple shapes simply as the union of the latter. Another use is in sculpting understood as removal of material: any grinding, milling, routing, or drilling operation that can be performed with physical machinery on physical materials can be simulated on the computer with the Boolean operation $x \wedge \neg y$ or $x - y$, which in set theory is set difference, remove the elements of y from those of x. Thus given two shapes one to be machined and the other the material to be removed, the result of machining the former to remove the latter is described simply as their set difference.

Boolean Searches

Search engine queries also employ Boolean logic. For this application, each web page on the Internet may be considered to be an "element" of a "set". The following examples use a syntax previously supported by Google:

- Doublequotes are used to combine whitespace-separated words into a single search term.

- Whitespace is used to specify logical AND, as it is the default operator for joining search terms:

"Search term 1" "Search term 2"

- The OR keyword is used for logical OR:

"Search term 1" OR "Search term 2"

- A prefixed minus sign is used for logical NOT:

"Search term 1" –"Search term 2"

Unary Algebra

A universal algebra $\langle A, \{f_i : i \in I\} \rangle$ with a family $\{f_i : i \in I\}$ of unary operations $f_i : A \to A$. An important example of a unary algebra arises from a group homomorphism $\phi : G \to S_A$ from an arbitrary group G into the group S_A of all permutations of a set A. Such a homomorphism is called an action of the group G on A. The definition, for each element $g \in G$, of a unary operation $f_g : A \to A$. as the permutation $\phi(g)$ in S_A corresponding to the element g under the homomorphism ϕ yields a unary algebra $\langle A, \{f\, g : g \in G\} \rangle$, in which

$$f_1(x) = x, \quad f_g\big(f_h(x)\big) = f_{gh}(x), \quad x \in A, \quad g, h \in G.$$

Every module over a ring carries a unary algebra structure. Every deterministic semi-automaton (cf. Automaton, algebraic theory of) with set S of states and input symbols $\alpha_1, \dots, \alpha_n$ may also be considered as a unary algebra $\langle S, f_1, \dots, f_n \rangle$, where $f_i(s) = \alpha_i s$ is the state onto which the state S is mapped by the action of the input symbol α_i.

A unary algebra with a single basic operation is called mono-unary, or a unar. An example of a unar is the Peano algebra $\langle p, f \rangle$, where $P = \{1, 2, \dots\}$ and $f(n) = n + 1$.

The identities of an arbitrary unary algebra can only be of the following types:

$$\mathrm{I}_1. f_{i_1} \dots f_{i_k}(x) = f_{j_1} \dots f_{j_l}(x),$$
$$\mathrm{II}_1. f_{i_1} \dots f_{i_k}(x) = f_{j_1} \dots f_{j_l}(x),$$
$$\mathrm{I}_2. f_{i_1} \dots f_{i_k}(x) = x,$$
$$\mathrm{II}_2. f_{i_1} \dots f_{i_k}(x) = y,$$
$$\mathrm{I}_3. x = x,$$
$$\mathrm{II}_3. x = y.$$

The identity II_2 is equivalent to II , being satisfied only by a 1-element algebra. A variety of unary algebras defined only by identities of the form I_1, I_2 or I_3 is said to be regular. There exists the following link between regular varieties of unary algebras and semi-groups.

Let V be a regular variety of unary algebras given by a set $\{f_i : i \in I\}$, $I = \emptyset$, of function symbols and a set Σ of identities. Each symbol f_i corresponds to an element α_i, and for every identity of the form I_1 from Σ one writes the defining relation

$$\alpha i_1 \dots \alpha i_k = \alpha_{j_1} \dots \alpha_{j_l}.$$

Let P be the semi-group with generators $\alpha_i, i \in I$, and the above defining relations, and let P^1 be the semi-group P with an identity e adjoined. For every relation of the form I_2 in Σ (if they are any) one writes the defining relation as $\alpha i_1 \ldots \alpha i_k = e$. The semi-group P_V obtained from P^1 by adjoining these defining relations is said to be associated with the variety V. There are many ways of characterizing this variety. If Σ contains only identities of the form I_1, then one may restrict oneself to the construction of P. By defining a unary operation $f_i(x) = xa_i$ in P_V one obtains a unary algebra $\langle P_V, \{f_i : i \in I\} \rangle$, which is a V-free algebra of rank 1. The group of all automorphisms of the unary algebra $\langle P_V, \{f_i : i \in I\} \rangle$ is isomorphic to the group P_V^* of invertible elements of the semi-group P_V.

References

- Randal Schwartz, Brian Foy, Tom Phoenix, Learning Perl, Publisher O'Reilly Media, Inc., 2011, ISBN 1449313140, 9781449313142

- "What is the following property of inequality called?". Stack Exchange. November 29, 2014. Retrieved 4 May 2018

- Universal Algebra: mathworld.wolfram.com, Retrieved 25 June 2019

- Mary Jane Sterling, Algebra II For Dummies, Publisher: John Wiley & Sons, 2006, ISBN 0471775819, 9780471775812, 384 pages

Algebraic Expressions

Algebraic expressions consist of variables, constants and other algebraic operations such as addition, subtraction, etc. There are three main types of algebraic expressions which are monomial expression, binomial expression and polynomial expression. This chapter has been carefully written to provide an easy understanding of these algebraic expressions.

An algebraic expression in mathematics is an expression which is made up of variables and constants along with algebraic operations (addition, subtraction, etc.). Expressions are made up of terms.

Example of Algebraic Expression

3x+4y -7, 4x − 10 etc.

It is to be noted that, unlike algebraic equation, an algebraic expression has no sides or equal to sign. Some of its examples include.

- 3x+4y -7

- 4x − 10

- $2x^2−3xy+5$

The Terminology used in Algebraic Expressions

In Algebra we work with Variable, Symbols or Letters whose value is unknown to us.

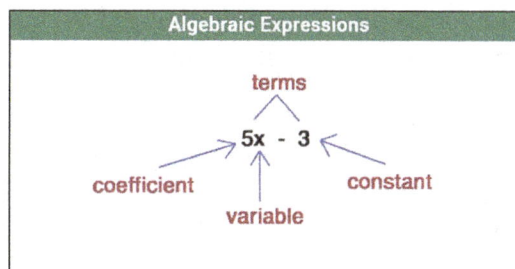

In the above expression (i.e. 5x − 3),

- x is a Variable, whose value is unknown to us which can take any value.

- 5 is known as the Coefficient of x, as it is a constant value used with the variable term and is well defined.

- 3 is the Constant value term which has a definite value.

The whole expression is known to be the Binomial term, as it has two unlikely terms.

Variables

In mathematics, a variable is a quantity that can change. Letters are used to represent these changing, unknown quantities.

Einstein's famous equation $E = MC^2$ uses the following variables:

- E for the amount of energy produced,
- M for the amount of mass used, and
- C^2 to represent the speed of light squared.

Variables, unknown quantities, are the opposite of constants, which are known, unchanging amounts.

Variables can be independent or dependent. Dependent and independent variables are commonly used in statistical studies and to control (to some degree) the outcomes of experiments. In the simple equation $y = 2x$, the letter x can be any real number. The value of y is completely dependent on the value chosen for x, and is always twice as much. Thus, x is the independent variable, and y is the dependent variable.

It is common for variables to play different roles in the same mathematical formula and names or qualifiers have been introduced to distinguish them. For example, the general cubic equation

$$ax^3 + bx^2 + cx + d = 0,$$

is interpreted as having five variables: four, a, b, c, d, which are taken to be given numbers and the fifth variable, x, is understood to be an *unknown* number. To distinguish them, the variable x is called *an unknown*, and the other variables are called *parameters* or *coefficients*, or sometimes *constants*, although this last terminology is incorrect for an equation and should be reserved for the function defined by the left-hand side of this equation.

In the context of functions, the term *variable* refers commonly to the arguments of the functions. This is typically the case in sentences like "function of a real variable", "x is the variable of the function $f: x \rightarrow f(x)$", "f is a function of the variable x" (meaning that the argument of the function is referred to by the variable x).

In the same context, variables that are independent of x define constant functions and are therefore called *constant*. For example, a *constant of integration* is an arbitrary constant function that is added to a particular antiderivative to obtain the other antiderivatives. Because the strong relationship between polynomials and polynomial function, the term "constant" is often used to denote the coefficients of a polynomial, which are constant functions of the indeterminates.

This use of "constant" as an abbreviation of "constant function" must be distinguished from the normal meaning of the word in mathematics. A constant, or mathematical constant is a well and unambiguously defined number or other mathematical object, as, for example, the numbers 0, 1, π and the identity element of a group.

Other specific names for variables are:

- An unknown is a variable in an equation which has to be solved for.

- An indeterminate is a symbol, commonly called variable, that appears in a polynomial or a formal power series. Formally speaking, an indeterminate is not a variable, but a constant in the polynomial ring or the ring of formal power series. However, because of the strong relationship between polynomials or power series and the functions that they define, many authors consider indeterminates as a special kind of variables.

- A parameter is a quantity (usually a number) which is a part of the input of a problem, and remains constant during the whole solution of this problem. For example, in mechanics the mass and the size of a solid body are *parameters* for the study of its movement. In computer science, *parameter* has a different meaning and denotes an argument of a function.

Free Variables and Bound Variables

- A random variable is a kind of variable that is used in probability theory and its applications.

It should be emphasized that all these denominations of variables are of semantic nature and that the way of computing with them (syntax) is the same for all.

Dependent and Independent Variables

In calculus and its application to physics and other sciences, it is rather common to consider a variable, say y, whose possible values depend on the value of another variable, say x. In mathematical terms, the *dependent* variable y represents the value of a function of x. To simplify formulas, it is often useful to use the same symbol for the dependent variable y and the function mapping x onto y. For example, the state of a physical system depends on measurable quantities such as the pressure, the temperature, the spatial position, and all these quantities vary when the system evolves, that is, they are function of the time. In the formulas describing the system, these quantities are represented by variables which are dependent on the time, and thus considered implicitly as functions of the time.

Therefore, in a formula, a dependent variable is a variable that is implicitly a function of another (or several other) variables. An independent variable is a variable that is not dependent.

The property of a variable to be dependent or independent depends often of the point of view and is not intrinsic. For example, in the notation $f(x, y, z)$, the three variables may be all independent and the notation represents a function of three variables. On the other hand, if y and z depend on x (are *dependent variables*) then the notation represents a function of the single *independent variable x*.

Examples:

If one defines a function f from the real numbers to the real numbers by,

$$f(x) = x^2 + \sin(x+4)$$

then x is a variable standing for the argument of the function being defined, which can be any real number. In the identity

$$\sum_{i=1}^{n} i = \frac{n^2 + n}{2}$$

the variable i is a summation variable which designates in turn each of the integers 1, 2, ..., n (it is also called index because its variation is over a discrete set of values) while n is a parameter (it does not vary within the formula).

In the theory of polynomials, a polynomial of degree 2 is generally denoted as $ax^2 + bx + c$, where a, b and c are called coefficients (they are assumed to be fixed, i.e., parameters of the problem considered) while x is called a variable. When studying this polynomial for its polynomial function this x stands for the function argument. When studying the polynomial as an object in itself, x is taken to be an indeterminate, and would often be written with a capital letter instead to indicate this status.

Constant

In mathematics, the adjective constant means non-varying. The noun constant may have two different meanings. It may refer to a fixed and well-defined number or other mathematical object. The term mathematical constant (and also physical constant) is sometimes used to distinguish this meaning from the other one. A constant may also refer to a constant function or its value (it is a common usage to identify them). Such a constant is commonly represented by a variable which does not depend on the main variable(s) of the studied problem. This is the case, for example, for a constant of integration which is an arbitrary constant function (not depending on the variable of integration) added to a particular antiderivative to get all the antiderivatives of the given function.

For example, a general quadratic function is commonly written as:

$$ax^2 + bx + c,$$

where a, b and c are constants (or parameters), while x is the variable, a placeholder for the argument of the function being studied. A more explicit way to denote this function is

$$x \mapsto ax^2 + bx + c,$$

which makes the function-argument status of x clear, and thereby implicitly the constant status of a, b and c. In this example a, b and c are coefficients of the polynomial. Since c occurs in a term that does not involve x, it is called the constant term of the polynomial and can be thought of as the coefficient of x^0; any polynomial term or expression of degree zero is a constant.

Constant Function

A constant may be used to define a constant function that ignores its arguments and always gives the same value. A constant function of a single variable, such as $f(x) = 5$, has a graph that is a horizontal straight line, parallel to the x-axis. Such a function always takes the same value because its argument does not appear in the expression defining the function.

Context-dependence

The context-dependent nature of the concept of "constant" can be seen in this example from elementary calculus:

$$\frac{d}{dx}2^x = \lim_{h \to 0}\frac{2^{x+h}-2^x}{h} = \lim_{h \to 0}2^x\frac{2^h-1}{h}$$

$$= 2^x \lim_{h \to 0}\frac{2^h-1}{h} \qquad \text{since } x \text{ is constant (i.e. does not depend on } h)$$

$$= 2^x \cdot \text{constant}, \qquad \text{where constant means not depending on } x.$$

"Constant" means not depending on some variable; not changing as that variable changes. In the first case above, it means not depending on h; in the second, it means not depending on x.

Notable Mathematical Constants

Some values occur frequently in mathematics and are conventionally denoted by a specific symbol. These standard symbols and their values are called mathematical constants. Examples include:

- 0 (zero).

- 1 (one), the natural number after zero.

- π (pi), the constant representing the ratio of a circle's circumference to its diameter, approximately equal to 3.14159265358979323846264643...

- e, approximately equal to 2.718281828459045235360287...

- i, the imaginary unit such that $i^2 = -1$.

- $\sqrt{2}$ (square root of 2), the length of the diagonal of a square with unit sides, approximately equal to 1.41421356237309504880168 8.

- φ (golden ratio), approximately equal to 1.61803398874989484820458 6, or algebraically, $\frac{1+\sqrt{5}}{2}$.

Constants in Calculus

In calculus, constants are treated in several different ways depending on the operation. For example, the derivative of a constant function is zero. This is because the derivative measures the rate of change of a function with respect to a variable, and since constants, by definition, do not change, their derivative is therefore zero. Conversely, when integrating a constant function, the constant is multiplied by the variable of integration. During the evaluation of a limit, the constant remains the same as it was before and after evaluation.

Integration of a function of one variable often involves a constant of integration. This arises because of the integral operator's nature as the inverse of the differential operator, meaning the aim of integration is to recover the original function before differentiation. The differential of a constant function is zero, as noted above, and the differential operator is a linear operator, so functions that only differ by a constant term have the same derivative. To acknowledge this, a constant of integration is added to an indefinite integral; this ensures that all possible solutions are included. The constant of integration is generally written as 'c' and represents a constant with a fixed but undefined value.

Examples:

If f is the constant function such that $f(x) = 72$ for every x then,

$$f'(x) = 0$$
$$\int f(x)dx = 72x + c.$$

Coefficient

In mathematics, a coefficient is a multiplicative factor in some term of a polynomial, a series, or any expression; it is usually a number, but may be any expression. In the latter case, the variables appearing in the coefficients are often called parameters, and must be clearly distinguished from the other variables.

For example, in

$$7x^2 - 3xy + 1.5 + y,$$

the first two terms respectively have the coefficients 7 and −3. The third term 1.5 is a constant co-efficient. The final term does not have any explicitly written coefficient, but is considered to have coefficient 1, since multiplying by that factor would not change the term.

Often coefficients are numbers as in this example, although they could be parameters of the problem or any expression in these parameters. In such a case one must clearly distinguish between symbols representing variables and symbols representing parameters. Following René Descartes, the variables are often denoted by x, y, ..., and the parameters by a, b, c, ..., but it is not always the case. For example, if y is considered as a parameter in the above expression, the coefficient of x is −3y, and the constant coefficient is 1.5 + y.

When one writes

$$ax^2 + bx + c,$$

it is generally supposed that x is the only variable and that a, b and c are parameters; thus the con-stant coefficient is c in this case.

Similarly, any polynomial in one variable x can be written as,

$$a_k x^k + \cdots + a_1 x^1 + a_0$$

for some positive integer k, where $a_k, .., a_1, a_0$ are coefficients; to allow this kind of expression in all cases one must allow introducing terms with 0 as coefficient. For the largest i with $a_i \neq 0$ (if any),

a_i is called the leading coefficient of the polynomial. So for example the leading coefficient of the polynomial

$$4x^5 + x^3 + 2x^2$$

is 4.

Some specific coefficients that occur frequently in mathematics have received a name. This is the case of the binomial coefficients, the coefficients which occur in the expanded form of $(x+y)^n$, and are tabulated in Pascal's triangle.

Linear Algebra

In linear algebra, the leading coefficient (also leading entry) of a row in a matrix is the first nonzero entry in that row. So, for example, given

$$M = \begin{pmatrix} 1 & 2 & 0 & 6 \\ 0 & 2 & 9 & 4 \\ 0 & 0 & 0 & 4 \\ 0 & 0 & 0 & 0 \end{pmatrix}.$$

The leading coefficient of the first row is 1; 2 is the leading coefficient of the second row; 4 is the leading coefficient of the third row, and the last row does not have a leading coefficient.

Though coefficients are frequently viewed as constants in elementary algebra, they can be variables more generally. For example, the coordinates $(x_1, x_2, \ldots, x_n)$ of a vector v in a vector space with basis $\{e_1, e_2, \ldots, e_n\}$, are the coefficients of the basis vectors in the expression

$$v = x_1 e_1 + x_2 e_2 + \cdots + x_n e_n.$$

Algebraic Expression Example

Simplify the given expressions by combining the like terms and write the type of Algebraic expression.

   i. $3xy^3 + 9x^2 y^3 + 8x^3 + 5y^3 x$

   ii. $7ab^2 c^2 + 2a^3 b^2 - 3abc - 5ab^2 c^2 - 2b^2 a^3 + 2ab$

  iii. $50x^3 - 20x + 83 + 21x^3 - 3x + 3 + 15x - 41x^3$

Solution:

Creating a table to find the solution:

S.no	Term	Simplification	Type of Expression
1	$3xy^3 + 9x^2 y^3 + 8x^3 + 5y^3 x$	$8xy^3 + 9x^2 y^3$	Binomial
2	$7ab^2 c^2 + 2a^3 b^2 - 3abc - 5ab^2 c^2 - 2b^2 a^3 + 2ab$	$2ab^2 c^2 - 3abc + 2ab$	Trinomial
3	$50x^3 - 20x + 83 + 21x^3 - 3x + 3 + 15x - 41x^3$	$3x^3$	Monomial

MONOMIAL

In mathematics, a monomial is, roughly speaking, a polynomial which has only one term. Two definitions of a monomial may be encountered:

- A monomial, also called power product, is a product of powers of variables with nonnegative integer exponents, or, in other words, a product of variables, possibly with repetitions. The constant 1 is a monomial, being equal to the empty product and x^0 for any variable x. If only a single variable x is considered, this means that a monomial is either 1 or a power x^n of x, with n a positive integer. If several variables are considered, say, x, y, z, then each can be given an exponent, so that any monomial is of the form $x^a y^b z^c$ with a, b, c non-negative integers (taking note that any exponent 0 makes the corresponding factor equal to 1).

- A monomial is a monomial in the first sense multiplied by a nonzero constant, called the coefficient of the monomial. A monomial in the first sense is a special case of a monomial in the second sense, where the coefficient is 1. For example, in this interpretation $-7x^5$ and $(3-4i)x^4 yz^{13}$ are monomials (in the second example, the variables are x, y, z, and the coefficient is a complex number).

In the context of Laurent polynomials and Laurent series, the exponents of a monomial may be negative, and in the context of Puiseux series, the exponents may be rational numbers.

Comparison of the Two Definitions

With either definition, the set of monomials is a subset of all polynomials that is closed under multiplication.

Both uses of this notion can be found, and in many cases the distinction is simply ignored, see for instance examples for the first and second meaning. In informal discussions the distinction is seldom important, and tendency is towards the broader second meaning. When studying the structure of polynomials however, one often definitely needs a notion with the first meaning. This is for instance the case when considering a monomial basis of a polynomial ring, or a monomial ordering of that basis. An argument in favor of the first meaning is also that no obvious other notion is available to designate these values (the term power product is in use, in particular when *monomial* is used with the first meaning, but it does not make the absence of constants clear either), while the notion term of a polynomial unambiguously coincides with the second meaning of monomial.

Monomial Basis

The most obvious fact about monomials (first meaning) is that any polynomial is a linear combination of them, so they form a basis of the vector space of all polynomials, called the *monomial basis* - a fact of constant implicit use in mathematics.

Number

The number of monomials of degree d in n variables is the number of multicombinations of d elements chosen among the n variables (a variable can be chosen more than once, but order does not matter), which is given by the multiset coefficient $\left(\!\!\left(\begin{array}{c} n \\ d \end{array}\right)\!\!\right)$. This expression can also be given in the form of a binomial coefficient, as a polynomial expression in d, or using a rising factorial power of $d + 1$:

$$\left(\!\!\left(\begin{array}{c} n \\ d \end{array}\right)\!\!\right) = \left(\begin{array}{c} n+d-1 \\ d \end{array}\right) = \left(\begin{array}{c} d+(n-1) \\ n-1 \end{array}\right) = \frac{(d+1)\times(d+2)\times\cdots\times(d+n-1)}{1\times2\times\cdots\times(n-1)} = \frac{1}{(n-1)!}(d+1)^{\overline{n-1}}.$$

The latter forms are particularly useful when one fixes the number of variables and lets the degree vary. From these expressions one sees that for fixed n, the number of monomials of degree d is a polynomial expression in d of degree $n-1$ with leading coefficient $\frac{1}{(n-1)!}$.

For example, the number of monomials in three variables ($n = 3$) of degree d is $\frac{1}{2}(d+1)^{\overline{2}} = \frac{1}{2}(d+1)(d+2)$; these numbers form the sequence 1, 3, 6, 10, 15, of triangular numbers.

The Hilbert series is a compact way to express the number of monomials of a given degree: the number of monomials of degree d in n variables is the coefficient of degree d of the formal power series expansion of,

$$\frac{1}{(1-t)^n}.$$

The number of monomials of degree at most d in n variables is $\left(\begin{array}{c} n+d \\ n \end{array}\right) = \left(\begin{array}{c} n+d \\ d \end{array}\right)$. This follows from the one-to-one correspondence between the monomials of degree d in $n+1$ variables and the monomials of degree at most d in n variables, which consists in substituting by 1 the extra variable.

Notation

Notation for monomials is constantly required in fields like partial differential equations. If the variables being used form an indexed family like x_1, x_2, x_3, then *multi-index notation* is helpful: if we write,

$$\alpha = (a,b,c)$$

we can define

$$x^\alpha = x_1^a x_2^b x_3^c$$

for compactness.

Degree

The degree of a monomial is defined as the sum of all the exponents of the variables, including the implicit exponents of 1 for the variables which appear without exponent; e.g., in the example of the previous section, the degree is $a+b+c$. The degree of xyz^2 is 1+1+2=4. The degree of a nonzero constant is 0. For example, the degree of -7 is 0.

The degree of a monomial is sometimes called order, mainly in the context of series. It is also called total degree when it is needed to distinguish it from the degree in one of the variables.

Monomial degree is fundamental to the theory of univariate and multivariate polynomials. Explicitly, it is used to define the degree of a polynomial and the notion of homogeneous polynomial, as well as for graded monomial orderings used in formulating and computing Gröbner bases. Implicitly, it is used in grouping the terms of a Taylor series in several variables.

Geometry

In algebraic geometry the varieties defined by monomial equations $x^{\alpha} = 0$ for some set of α have special properties of homogeneity. This can be phrased in the language of algebraic groups, in terms of the existence of a group action of an algebraic torus (equivalently by a multiplicative group of diagonal matrices). This area is studied under the name of *torus embeddings*.

BINOMIAL

In algebra, a binomial is a polynomial that is the sum of two terms, each of which is a monomial. It is the simplest kind of polynomial after the monomials.

A binomial is a polynomial which is the sum of two monomials. A binomial in a single indeterminate (also known as a univariate binomial) can be written in the form

$$ax^m - bx^n,$$

where a and b are numbers, and m and n are distinct nonnegative integers and x is a symbol which is called an indeterminate or, for historical reasons, a variable. In the context of Laurent polynomials, a *Laurent binomial*, often simply called a *binomial*, is similarly defined, but the exponents m and n may be negative.

More generally, a binomial may be written as:

$$ax_1^{n_1} \cdots x_i^{n_i} - bx_1^{m_1} \cdots x_i^{m_i}$$

Some examples of binomials are:

$$3x - 2x^2$$

$$xy + yx^2$$

$$0.9x^3 + \pi y^2$$

Operations on Simple Binomials

- The binomial $x^2 - y^2$ can be factored as the product of two other binomials:

$$x^2 - y^2 = (x + y)(x - y).$$

This is a special case of the more general formula:

$$x^{n+1} - y^{n+1} = (x-y)\sum_{k=0}^{n} x^k y^{n-k}.$$

When working over the complex numbers, this can also be extended to:

$$x^2 + y^2 = x^2 - (iy)^2 = (x-iy)(x+iy).$$

- The product of a pair of linear binomials $(ax + b)$ and $(cx + d)$ is a trinomial:

$$(ax+b)(cx+d) = acx^2 + (ad+bc)x + bd.$$

- A binomial raised to the nth power, represented as $(x + y)^n$ can be expanded by means of the binomial theorem or, equivalently, using Pascal's triangle. For example, the square $(x + y)^2$ of the binomial $(x + y)$ is equal to the sum of the squares of the two terms and twice the product of the terms, that is:

$$(x+y)^2 = x^2 + 2xy + y^2.$$

The numbers (1, 2, 1) appearing as multipliers for the terms in this expansion are binomial coefficients two rows down from the top of Pascal's triangle. The expansion of the nth power uses the numbers n rows down from the top of the triangle.

- An application of above formula for the square of a binomial is the "(m, n)-formula" for generating Pythagorean triples:

For $m < n$, let $a = n^2 - m^2$, $b = 2mn$, and $c = n^2 + m^2$; then $a^2 + b^2 = c^2$.

- Binomials that are sums or differences of cubes can be factored into lower-order polynomials as follows:

$$x^3 + y^3 = (x+y)(x^2 - xy + y^2)$$

$$x^3 - y^3 = (x-y)(x^2 + xy + y^2)$$

POLYNOMIAL

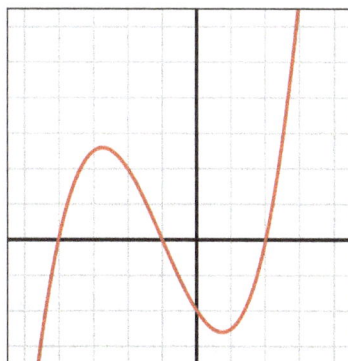

The graph of a polynomial function of degree 3.

In mathematics, a polynomial is an expression consisting of variables (also called indeterminates) and coefficients, that involves only the operations of addition, subtraction, multiplication, and non-negative integer exponents of variables. An example of a polynomial of a single indeterminate, x, is $x^2 - 4x + 7$. An example in three variables is $x^3 + 2xyz^2 - yz + 1$.

Polynomials appear in many areas of mathematics and science. For example, they are used to form polynomial equations, which encode a wide range of problems, from elementary word problems to complicated scientific problems; they are used to define polynomial functions, which appear in settings ranging from basic chemistry and physics to economics and social science; they are used in calculus and numerical analysis to approximate other functions. In advanced mathematics, polynomials are used to construct polynomial rings and algebraic varieties, central concepts in algebra and algebraic geometry.

A polynomial is an expression that can be built from constants and symbols called indeterminates or variables by means of addition, multiplication and exponentiation to a non-negative integer power. Two such expressions that may be transformed, one to the other, by applying the usual properties of commutativity, associativity and distributivity of addition and multiplication are considered as defining the same polynomial.

A polynomial in a single indeterminate x can always be written (or rewritten) in the form

$$a_n x^n + a_{n-1} x^{n-1} + \cdots + a_2 x^2 + a_1 x + a_0,$$

where $a_0, \ldots, a_n$ are constants and x is the indeterminate. The word "indeterminate" means that x represents no particular value, although any value may be substituted for it. The mapping that associates the result of this substitution to the substituted value is a function, called a *polynomial function*.

This can be expressed more concisely by using summation notation:

$$\sum_{k=0}^{n} a_k x^k$$

That is, a polynomial can either be zero or can be written as the sum of a finite number of non-zero terms. Each term consists of the product of a number—called the coefficient of the term—and a finite number of indeterminates, raised to nonnegative integer powers.

The x occurring in a polynomial is commonly called either a *variable* or an *indeterminate*. When the polynomial is considered as an expression, x is a fixed symbol which does not have any value (its value is "indeterminate"). However, when one considers the function defined by the polynomial, then x represents the argument of the function, and is therefore called a "variable". Many authors use these two words interchangeably.

It is common to use uppercase letters for indeterminates and corresponding lowercase letters for the variables (or arguments) of the associated function.

A polynomial P in the indeterminate x is commonly denoted either as P or as $P(x)$. Formally, the name of the polynomial is P, not $P(x)$, but the use of the functional notation $P(x)$ date from the time where the distinction between a polynomial and the associated function was unclear. Moreover, the functional notation is often useful for specifying, in a single phrase, a polynomial and its indeterminate. For example, "let $P(x)$ be a polynomial" is a shorthand for "let P be a polynomial in the indeterminate x". On the other hand, when it is not necessary to emphasize the name of the indeterminate, many formulas are much simpler and easier to read if the name(s) of the indeterminate(s) do not appear at each occurrence of the polynomial.

The ambiguity of having two notations for a single mathematical object may be formally resolved by considering the general meaning of the functional notation for polynomials. If a denotes a number, a variable, another polynomial, or, more generally any expression, then $P(a)$ denotes, by convention, the result of substituting a for x in P. Thus, the polynomial P defines the function

$$a \mapsto P(a),$$

which is the *polynomial function* associated to P. Frequently, when using this notation, one supposes that a is a number. However one may use it over any domain where addition and multiplication are defined (that is, any ring). In particular, if a is a polynomial then $P(a)$ is also a polynomial.

More specifically, when a is the indeterminate x, then the image of x by this function is the polynomial P itself (substituting x to x does not change anything). In other words,

$$P(x) = P,$$

which justifies formally the existence of two notations for the same polynomial.

Classification

The exponent on an indeterminate in a term is called the degree of that indeterminate in that term; the degree of the term is the sum of the degrees of the indeterminates in that term, and the degree of a polynomial is the largest degree of any one term with nonzero coefficient. Because $x = x^1$, the degree of an indeterminate without a written exponent is one.

A term with no indeterminates and a polynomial with no indeterminates are called, respectively, a constant term and a constant polynomial. The degree of a constant term and of a nonzero constant polynomial is 0. The degree of the zero polynomial, 0, (which has no terms at all) is generally treated as not defined.

For example:

$$-5x^2 y$$

is a term. The coefficient is -5, the indeterminates are x and y, the degree of x is two, while the degree of y is one. The degree of the entire term is the sum of the degrees of each indeterminate in it, so in this example the degree is $2 + 1 = 3$.

Forming a sum of several terms produces a polynomial. For example, the following is a polynomial:

$$\underbrace{3x}_{\text{term}\ 1}\ \underbrace{-5x}_{\text{term}\ 2}\ \underbrace{+4}_{\text{term}}.$$

It consists of three terms: the first is degree two, the second is degree one, and the third is degree zero.

Polynomials of small degree have been given specific names. A polynomial of degree zero is a *constant polynomial* or simply a *constant*. Polynomials of degree one, two or three are respectively *linear polynomials, quadratic polynomials* and *cubic polynomials*. For higher degrees the specific names are not commonly used, although *quartic polynomial* (for degree four) and *quintic polynomial* (for degree five) are sometimes used. The names for the degrees may be applied to the polynomial or to its terms. For example, in $x^2 + 2x + 1$ the term $2x$ is a linear term in a quadratic polynomial.

The polynomial 0, which may be considered to have no terms at all, is called the zero polynomial. Unlike other constant polynomials, its degree is not zero. Rather the degree of the zero polynomial is either left explicitly undefined, or defined as negative (either -1 or $-\infty$). These conventions are useful when defining Euclidean division of polynomials. The zero polynomial is also unique in that it is the only polynomial in one indeterminate having an infinite number of roots. The graph of the zero polynomial, $f(x) = 0$, is the X-axis.

In the case of polynomials in more than one indeterminate, a polynomial is called *homogeneous* of degree n if *all* its non-zero terms have degree n. The zero polynomial is homogeneous, and, as homogeneous polynomial, its degree is undefined. For example, $x^3y^2 + 7x^2y^3 - 3x^5$ is homogeneous of degree 5.

The commutative law of addition can be used to rearrange terms into any preferred order. In polynomials with one indeterminate, the terms are usually ordered according to degree, either in "descending powers of x", with the term of largest degree first, or in "ascending powers of x". The polynomial in the example above is written in descending powers of x. The first term has coefficient 3, indeterminate x, and exponent 2. In the second term, the coefficient is -5. The third term is a constant. Because the *degree* of a non-zero polynomial is the largest degree of any one term, this polynomial has degree two.

Two terms with the same indeterminates raised to the same powers are called "similar terms" or "like terms", and they can be combined, using the distributive law, into a single term whose coefficient is the sum of the coefficients of the terms that were combined. It may happen that this makes the coefficient 0. Polynomials can be classified by the number of terms with nonzero coefficients, so that a one-term polynomial is called a monomial, a two-term polynomial is called a binomial, and a three-term polynomial is called a *trinomial*. The term "quadrinomial" is occasionally used for a four-term polynomial.

A real polynomial is a polynomial with real coefficients. When it is used to define a function, the domain is not so restricted. However, a real polynomial function is a function from the reals to the reals that is defined by a real polynomial. Similarly, an integer polynomial is a polynomial with integer coefficients, and a complex polynomial is a polynomial with complex coefficients.

A polynomial in one indeterminate is called a *univariate polynomial*, a polynomial in more than one indeterminate is called a multivariate polynomial. A polynomial with two indeterminates is called a bivariate polynomial. These notions refer more to the kind of polynomials one is generally working with than to individual polynomials; for instance when working with univariate polynomials one does not exclude constant polynomials (which may result, for instance, from the subtraction of non-constant polynomials), although strictly speaking constant polynomials do not contain any indeterminates at all. It is possible to further classify multivariate polynomials as *bivariate*, *trivariate*, and so on, according to the maximum number of indeterminates allowed. Again, so that the set of objects under consideration be closed under subtraction, a study of trivariate polynomials usually allows bivariate polynomials, and so on. It is common, also, to say simply "polynomials in x, y, and z", listing the indeterminates allowed.

The *evaluation of a polynomial* consists of substituting a numerical value to each indeterminate and carrying out the indicated multiplications and additions. For polynomials in one indeterminate, the evaluation is usually more efficient (lower number of arithmetic operations to perform) using Horner's method:

$$(((\cdots((a_nx+a_{n-1})x+a_{n-2})x+\cdots+a_3)x+a_2)x+a_1)x+a_0.$$

Arithmetic

Polynomials can be added using the associative law of addition (grouping all their terms together into a single sum), possibly followed by reordering, and combining of like terms. For example, if

$$P = 3x^2 - 2x + 5xy - 2$$
$$Q = -3x^2 + 3x + 4y^2 + 8$$

then

$$P + Q = 3x^2 - 2x + 5xy - 2 - 3x^2 + 3x + 4y^2 + 8$$

which can be simplified to,

$$P + Q = x + 5xy + 4y^2 + 6$$

To work out the product of two polynomials into a sum of terms, the distributive law is repeatedly applied, which results in each term of one polynomial being multiplied by every term of the other. For example, if

$$P = 2x + 3y + 5$$
$$Q = 2x + 5y + xy + 1$$

then

$$
\begin{aligned}
PQ \quad = \quad & (2x\cdot2x) \;+\; (2x\cdot5y) \;+\; (2x\cdot xy) \;+\; (2x\cdot1) \\
+ \;& (3y\cdot2x) \;+\; (3y\cdot5y) \;+\; (3y\cdot xy) \;+\; (3y\cdot1) \\
+ \;& (5\cdot2x) \;+\; (5\cdot5y) \;+\; (5\cdot xy) \;+\; (5\cdot1)
\end{aligned}
$$

which can be simplified to,

$$PQ = 4x^2 + 21xy + 2x^2y + 12x + 15y^2 + 3xy^2 + 28y + 5.$$

Polynomial evaluation can be used to compute the remainder of polynomial division by a polynomial of degree one, because the remainder of the division of $f(x)$ by $(x - a)$ is $f(a)$. This is more efficient than the usual algorithm of division when the quotient is not needed.

- A sum of polynomials is a polynomial.

- A product of polynomials is a polynomial.

- A composition of two polynomials is a polynomial, which is obtained by substituting a variable of the first polynomial by the second polynomial.

- The derivative of the polynomial $a_n x^n + a_{n-1} x^{n-1} + \ldots + a_2 x^2 + a_1 x + a_0$ is the polynomial $n a_n x^{n-1} + (n-1) a_{n-1} x^{n-2} + \ldots + 2 a_2 x + a_1$. If the set of the coefficients does not contain the integers (for example if the coefficients are integers modulo some prime number p), then $k a_k$ should be interpreted as the sum of a_k with itself, k times. For example, over the integers modulo p, the derivative of the polynomial $x^p + 1$ is the polynomial 0.

- A primitive integral or antiderivative of the polynomial $a_n x^n + a_{n-1} x^{n-1} + \cdots + a_2 x^2 + a_1 x + a_0$ is the polynomial $a_n x^{n+1}/(n+1) + a_{n-1} x^n/n + \cdots + a_2 x^3/3 + a_1 x^2/2 + a_0 x + c$, where c is an arbitrary constant. For instance, the antiderivatives of $x^2 + 1$ have the form $1/3 x^3 + x + c$.

As for the integers, two kinds of divisions are considered for the polynomials. The *Euclidean division of polynomials* that generalizes the Euclidean division of the integers. It results in two polynomials, a *quotient* and a *remainder* that are characterized by the following property of the polynomials: given two polynomials a and b such that $b \neq 0$, there exists a unique pair of polynomials, q, the quotient, and r, the remainder, such that $a = b\,q + r$ and degree(r) < degree(b) (here the polynomial zero is supposed to have a negative degree). By hand as well as with a computer, this division can be computed by the polynomial long division algorithm.

All polynomials with coefficients in a unique factorization domain (for example, the integers or a field) also have a factored form in which the polynomial is written as a product of irreducible polynomials and a constant. This factored form is unique up to the order of the factors and their multiplication by an invertible constant. In the case of the field of complex numbers, the irreducible factors are linear. Over the real numbers, they have the degree either one or two. Over the integers and the rational numbers the irreducible factors may have any degree. For example, the factored form of

$$5x^3 - 5$$

is

$$5(x-1)\left(x^2 + x + 1\right)$$

over the integers and the reals and

$$5(x-1)\left(x + \frac{1 + i\sqrt{3}}{2}\right)\left(x + \frac{1 - i\sqrt{3}}{2}\right)$$

over the complex numbers.

The computation of the factored form, called *factorization* is, in general, too difficult to be done by hand-written computation. However, efficient polynomial factorization algorithms are available in most computer algebra systems.

A formal quotient of polynomials, that is, an algebraic fraction wherein the numerator and denominator are polynomials, is called a "rational expression" or "rational fraction" and is not, in general, a polynomial. Division of a polynomial by a number, however, yields another polynomial. For example, $x^3/12$ is considered a valid term in a polynomial (and a polynomial by itself) because it is equivalent to $(1/12)x^3$ and $1/12$ is just a constant. When this expression is used as a term, its coefficient is therefore $1/12$. For similar reasons, if complex coefficients are allowed, one may have a single term like $(2 + 3i)\, x^3$; even though it looks like it should be expanded to two terms, the complex number $2 + 3i$ is one complex number, and is the coefficient of that term. The expression $1/(x^2 + 1)$ is not a polynomial because it includes division by a non-constant polynomial. The expression $(5 + y)^x$ is not a polynomial, because it contains an indeterminate used as exponent.

Because subtraction can be replaced by addition of the opposite quantity, and because positive integer exponents can be replaced by repeated multiplication, all polynomials can be constructed from constants and indeterminates using only addition and multiplication.

Polynomial Functions

A *polynomial function* is a function that can be defined by evaluating a polynomial. More precisely, a function f of one argument from a given domain is a polynomial function if there exists a polynomial,

$$a_n x^n + a_{n-1} x^{n-1} + \cdots + a_2 x^2 + a_1 x + a_0$$

that evaluates to $f(x)$ for all x in the domain of f (here, n is a non-negative integer and a_0, a_1, a_2, ..., a_n are constant coefficients).

Generally, unless otherwise specified, polynomial functions have complex coefficients, arguments, and values. In particular, a polynomial, restricted to have real coefficients, defines a function from the complex numbers to the complex numbers. If the domain of this function is also restricted to the reals, the resulting function maps reals to reals.

For example, the function f, defined by

$$f(x) = x^3 - x,$$

is a polynomial function of one variable. Polynomial functions of several variables are similarly defined, using polynomials in more than one indeterminate, as in

$$f(x, y) = 2x^3 + 4x^2 y + xy^5 + y^2 - 7.$$

According to the definition of polynomial functions, there may be expressions that obviously are not polynomials but nevertheless define polynomial functions. An example is the expression $\left(\sqrt{1 - x^2}\right)^2$, which takes the same values as the polynomial $1 - x^2$ on the interval $[-1,1]$, and thus both expressions define the same polynomial function on this interval.

Every polynomial function is continuous, smooth, and entire.

Graphs

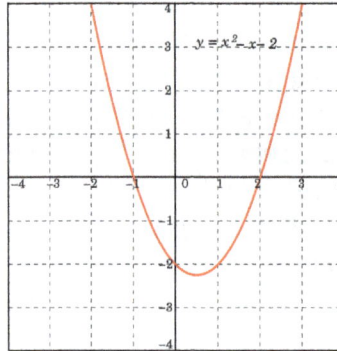

Polynomial of degree 2:

$f(x) = x^2 - x - 2$

$= (x + 1)(x - 2)$

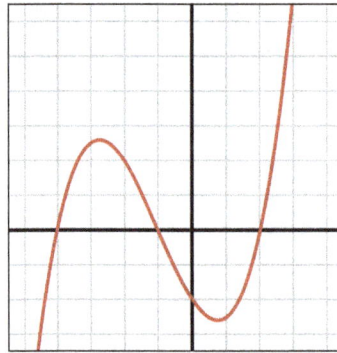

Polynomial of degree 3:

$f(x) = x^3/4 + 3x^2/4 - 3x/2 - 2$

$= 1/4 \ (x + 4)(x + 1)(x - 2)$

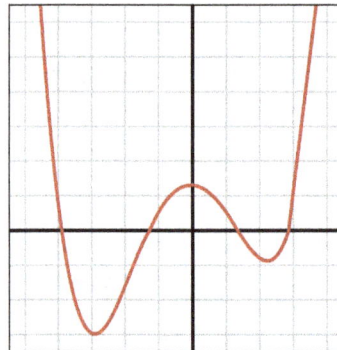

Polynomial of degree 4:

$f(x) = 1/14 \ (x + 4)(x + 1)(x - 1)(x - 3)$

$+ 0.5$

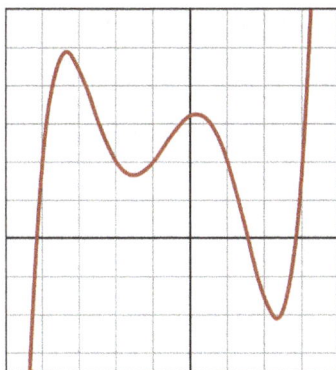

Polynomial of degree 5:

$f(x) = 1/20\ (x + 4)(x + 2)(x + 1)(x - 1)$

$(x - 3) + 2$

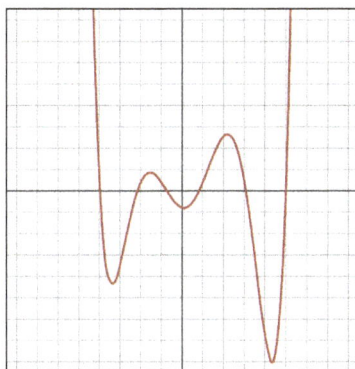

Polynomial of degree 6:

$f(x) = 1/100\ (x^6 - 2x^5 - 26x^4 + 28x^3$

$+ 145x^2 - 26x - 80)$

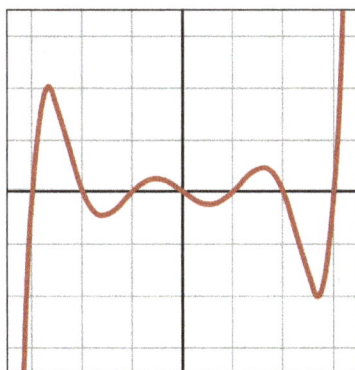

Polynomial of degree 7:

$f(x) = (x - 3)(x - 2)(x - 1)(x)(x + 1)(x + 2)$

$(x + 3)$

A polynomial function in one real variable can be represented by a graph.

- The graph of the zero polynomial

  $f(x) = 0$

  is the x-axis.

- The graph of a degree 0 polynomial

  $f(x) = a_0$, where $a_0 \neq 0$,

  is a horizontal line with y-intercept a_0.

- The graph of a degree 1 polynomial (or linear function)

  $f(x) = a_0 + a_1 x$, where $a_1 \neq 0$,

  is an oblique line with y-intercept a_0 and slope a_1.

- The graph of a degree 2 polynomial

  $f(x) = a_0 + a_1 x + a_2 x^2$, where $a_2 \neq 0$

  is a parabola.

- The graph of a degree 3 polynomial

  $f(x) = a_0 + a_1 x + a_2 x^2 + a_3 x^3$, where $a_3 \neq 0$

  is a cubic curve.

- The graph of any polynomial with degree 2 or greater

  $f(x) = a_0 + a_1 x + a_2 x^2 + \ldots + a_n x^n$, where $a_n \neq 0$ and $n \geq 2$

  is a continuous non-linear curve.

A non-constant polynomial function tends to infinity when the variable increases indefinitely (in absolute value). If the degree is higher than one, the graph does not have any asymptote. It has two parabolic branches with vertical direction (one branch for positive x and one for negative x).

Polynomial graphs are analyzed in calculus using intercepts, slopes, concavity, and end behavior.

Equations

A *polynomial equation*, also called *algebraic equation*, is an equation of the form

$$a_n x^n + a_{n-1} x^{n-1} + \cdots + a_2 x^2 + a_1 x + a_0 = 0.$$

For example,

$$3x^2 + 4x - 5 = 0$$

is a polynomial equation.

When considering equations, the indeterminates (variables) of polynomials are also called unknowns, and the *solutions* are the possible values of the unknowns for which the equality is true (in general more than one solution may exist). A polynomial equation stands in contrast to a *polynomial identity* like $(x + y)(x - y) = x^2 - y^2$, where both expressions represent the same polynomial in different forms, and as a consequence any evaluation of both members gives a valid equality.

In elementary algebra, methods such as the quadratic formula are taught for solving all first degree and second degree polynomial equations in one variable. There are also formulas for the cubic and quartic equations. For higher degrees, the Abel–Ruffini theorem asserts that there can not exist a general formula in radicals. However, root-finding algorithms may be used to find numerical approximations of the roots of a polynomial expression of any degree.

The number of real solutions of a polynomial equation with real coefficients may not exceed the degree, and equals the degree when the complex solutions are counted with their multiplicity. This fact is called the fundamental theorem of algebra.

Solving Equations

Every polynomial P in x defines a function $x \mapsto P(x)$, called the *polynomial function* associated to P; the equation $P(x) = 0$ is the *polynomial equation* associated to P. The solutions of this equation are called the *roots* of the polynomial, or the zeros of the associated function (they correspond to the points where the graph of the function meets the x-axis).

A number a is a root of a polynomial P if and only if the linear polynomial $x - a$ divides P, that is if there is another polynomial Q such that $P = (x - a) Q$. It may happen that $x - a$ divides P more than once: if $(x - a)^2$ divides P then a is called a *multiple root* of P, and otherwise a is called a *simple root* of P. If P is a nonzero polynomial, there is a highest power m such that $(x - a)^m$ divides P, which is called the *multiplicity* of the root a in P. When P is the zero polynomial, the corresponding polynomial equation is trivial, and this case is usually excluded when considering roots, as, with the above definitions, every number is a root of the zero polynomial, with an undefined multiplicity. With this exception made, the number of roots of P, even counted with their respective multiplicities, cannot exceed the degree of P. The relation between the coefficients of a polynomial and its roots is described by Vieta's formulas.

Some polynomials, such as $x^2 + 1$, do not have any roots among the real numbers. If, however, the set of accepted solutions is expanded to the complex numbers, every non-constant polynomial has at least one root; this is the fundamental theorem of algebra. By successively dividing out factors $x - a$, one sees that any polynomial with complex coefficients can be written as a constant (its leading coefficient) times a product of such polynomial factors of degree 1; as a consequence, the number of (complex) roots counted with their multiplicities is exactly equal to the degree of the polynomial.

There may be several meanings of "solving an equation". One may want to express the solutions as explicit numbers; for example, the unique solution of $2x - 1 = 0$ is 1/2. Unfortunately, this is, in general, impossible for equations of degree greater than one, and, since the ancient times, mathematicians have searched to express the solutions as algebraic expression; for example the golden ratio $(1 + \sqrt{5})/2$ is the unique positive solution of $x^2 - x - 1 = 0$. In the ancient times, they succeeded only for degrees one and two. For quadratic equations, the quadratic formula provides such expressions of the solutions. Since the 16th century, similar formulas (using cube roots in addition to square roots), but much more complicated are known for equations of degree three and four. But formulas for degree 5 and higher eluded researchers for several centuries. In 1824, Niels Henrik Abel proved the striking result that there are equations of degree 5 whose solutions cannot be expressed by a (finite) formula, involving only arithmetic operations and radicals. In 1830, Évariste Galois proved that most equations of degree higher than four cannot be solved by radicals, and showed that for each equation, one may decide whether it is solvable by radicals, and, if it is, solve it. This result marked the start of Galois theory and group theory, two important branches of modern algebra. Galois himself noted that the computations implied by his method were impracticable. Nevertheless, formulas for solvable equations of degrees 5 and 6 have been published.

When there is no algebraic expression for the roots, and when such an algebraic expression exists but is too complicated to be useful, the unique way of solving is to compute numerical approximations of the solutions. There are many methods for that; some are restricted to polynomials and others may apply to any continuous function. The most efficient algorithms allow solving easily (on a computer) polynomial equations of degree higher than 1,000.

For polynomials in more than one indeterminate, the combinations of values for the variables for which the polynomial function takes the value zero are generally called *zeros* instead of "roots". The study of the sets of zeros of polynomials is the object of algebraic geometry. For a set of polynomial equations in several unknowns, there are algorithms to decide whether they have a finite number of complex solutions, and, if this number is finite, for computing the solutions.

The special case where all the polynomials are of degree one is called a system of linear equations, for which another range of different solution methods exist, including the classical Gaussian elimination.

A polynomial equation for which one is interested only in the solutions which are integers is called a Diophantine equation. Solving Diophantine equations is generally a very hard task. It has been proved that there cannot be any general algorithm for solving them, and even for deciding whether the set of solutions is empty. Some of the most famous problems that have been solved during the fifty last years are related to Diophantine equations, such as Fermat's Last Theorem.

Generalizations

There are several generalizations of the concept of polynomials.

Trigonometric Polynomials

A trigonometric polynomial is a finite linear combination of functions $\sin(nx)$ and $\cos(nx)$ with n taking on the values of one or more natural numbers. The coefficients may be taken as real numbers, for real-valued functions.

If $\sin(nx)$ and $\cos(nx)$ are expanded in terms of $\sin(x)$ and $\cos(x)$, a trigonometric polynomial becomes a polynomial in the two variables $\sin(x)$ and $\cos(x)$ (using List of trigonometric identities#-Multiple-angle formulae). Conversely, every polynomial in $\sin(x)$ and $\cos(x)$ may be converted, with Product-to-sum identities, into a linear combination of functions $\sin(nx)$ and $\cos(nx)$. This equivalence explains why linear combinations are called polynomials.

For complex coefficients, there is no difference between such a function and a finite Fourier series.

Trigonometric polynomials are widely used, for example in trigonometric interpolation applied to the interpolation of periodic functions. They are used also in the discrete Fourier transform.

Matrix Polynomials

A matrix polynomial is a polynomial with square matrices as variables. Given an ordinary, scalar-valued polynomial

$$P(x) = \sum_{i=0}^{n} a_i x^i = a_0 + a_1 x + a_2 x^2 + \cdots + a_n x^n,$$

this polynomial evaluated at a matrix A is

$$P(A) = \sum_{i=0}^{n} a_i A^i = a_0 I + a_1 A + a_2 A^2 + \cdots + a_n A^n,$$

where I is the identity matrix.

A matrix polynomial equation is an equality between two matrix polynomials, which holds for the specific matrices in question. A matrix polynomial identity is a matrix polynomial equation which holds for all matrices A in a specified matrix ring $M_n(R)$.

Laurent Polynomials

Laurent polynomials are like polynomials, but allow negative powers of the variable(s) to occur.

Rational Functions

A rational fraction is the quotient (algebraic fraction) of two polynomials. Any algebraic expression that can be rewritten as a rational fraction is a rational function.

While polynomial functions are defined for all values of the variables, a rational function is defined only for the values of the variables for which the denominator is not zero.

The rational fractions include the Laurent polynomials, but do not limit denominators to powers of an indeterminate.

Power Series

Formal power series are like polynomials, but allow infinitely many non-zero terms to occur, so that they do not have finite degree. Unlike polynomials they cannot in general be explicitly and fully written down (just like irrational numbers cannot), but the rules for manipulating their terms are the same as for polynomials. Non-formal power series also generalize polynomials, but the multiplication of two power series may not converge.

Other Examples

- A bivariate polynomial where the second variable is substituted by an exponential function applied to the first variable, for example $P(x, e^x)$, may be called an exponential polynomial.

Applications

Calculus

The simple structure of polynomial functions makes them quite useful in analyzing general functions using polynomial approximations. An important example in calculus is Taylor's theorem, which roughly states that every differentiable function locally looks like a polynomial function, and the Stone–Weierstrass theorem, which states that every continuous function defined on a compact interval of the real axis can be approximated on the whole interval as closely as desired by a polynomial function.

Calculating derivatives and integrals of polynomial functions is particularly simple. For the polynomial function,

$$\sum_{i=0}^{n} a_i x^i$$

the derivative with respect to x is,

$$\sum_{i=1}^{n} a_i i x^{i-1}$$

and the indefinite integral is:

$$\sum_{i=0}^{n} \frac{a_i}{i+1} x^{i+1} + c.$$

Abstract Algebra

In abstract algebra, one distinguishes between *polynomials* and *polynomial functions*. A *polynomial* f in one indeterminate x over a ring R is defined as a formal expression of the form:

$$f = a_n x^n + a_{n-1} x^{n-1} + \cdots + a_1 x^1 + a_0 x^0$$

where n is a natural number, the coefficients $a_0, \ldots, a_n$ are elements of R, and x is a formal symbol, whose powers x^i are just placeholders for the corresponding coefficients a_i, so that the given formal expression is just a way to encode the sequence $(a_0, a_1, \ldots)$, where there is an n such that $a_i = 0$ for all $i > n$. Two polynomials sharing the same value of n are considered equal if and only if the sequences of their coefficients are equal; furthermore any polynomial is equal to any polynomial with greater value of n obtained from it by adding terms in front whose coefficient is zero. These polynomials can be added by simply adding corresponding coefficients (the rule for extending by terms with zero coefficients can be used to make sure such coefficients exist). Thus each polynomial is actually equal to the sum of the terms used in its formal expression, if such a term $a_i x^i$ is interpreted as a polynomial that has zero coefficients at all powers of x other than

x^i. Then to define multiplication, it suffices by the distributive law to describe the product of any two such terms, which is given by the rule $ax^k \, bx^l = abx^{k+l}$ for all elements a, b of the ring R and all natural numbers k and l.

Thus the set of all polynomials with coefficients in the ring R forms itself a ring, the *ring of polynomials* over R, which is denoted by $R[x]$. The map from R to $R[x]$ sending r to rx^o is an injective homomorphism of rings, by which R is viewed as a subring of $R[x]$. If R is commutative, then $R[x]$ is an algebra over R.

One can think of the ring $R[x]$ as arising from R by adding one new element x to R, and extending in a minimal way to a ring in which x satisfies no other relations than the obligatory ones, plus commutation with all elements of R (that is $xr = rx$). To do this, one must add all powers of x and their linear combinations as well.

Formation of the polynomial ring, together with forming factor rings by factoring out ideals, are important tools for constructing new rings out of known ones. For instance, the ring (in fact field) of complex numbers, which can be constructed from the polynomial ring $R[x]$ over the real numbers by factoring out the ideal of multiples of the polynomial $x^2 + 1$. Another example is the construction of finite fields, which proceeds similarly, starting out with the field of integers modulo some prime number as the coefficient ring R.

If R is commutative, then one can associate to every polynomial P in $R[x]$, a *polynomial function f* with domain and range equal to R (more generally one can take domain and range to be the same unital associative algebra over R). One obtains the value $f(r)$ by substitution of the value r for the symbol x in P. One reason to distinguish between polynomials and polynomial functions is that over some rings different polynomials may give rise to the same polynomial function. This is not the case when R is the real or complex numbers, whence the two concepts are not always distinguished in analysis. An even more important reason to distinguish between polynomials and polynomial functions is that many operations on polynomials (like Euclidean division) require looking at what a polynomial is composed of as an expression rather than evaluating it at some constant value for x.

Divisibility

In commutative algebra, one major focus of study is *divisibility* among polynomials. If R is an integral domain and f and g are polynomials in $R[x]$, it is said that f *divides* g or f is a divisor of g if there exists a polynomial q in $R[x]$ such that $fq = g$. One can show that every zero gives rise to a linear divisor, or more formally, if f is a polynomial in $R[x]$ and r is an element of R such that $f(r) = 0$, then the polynomial $(x - r)$ divides f. The converse is also true. The quotient can be computed using the polynomial long division.

If F is a field and f and g are polynomials in $F[x]$ with $g \neq 0$, then there exist unique polynomials q and r in $F[x]$ with,

$$f = qg + r$$

and such that the degree of r is smaller than the degree of g (using the convention that the polynomial 0 has a negative degree). The polynomials q and r are uniquely determined by f and g. This is called *Euclidean division, division with remainder* or *polynomial long division* and shows that the ring $F[x]$ is a Euclidean domain.

Analogously, *prime polynomials* (more correctly, *irreducible polynomials*) can be defined as *non-zero polynomials which cannot be factorized into the product of two non-constant polynomials*. In the case of coefficients in a ring, *"non-constant"* must be replaced by *"non-constant or non-unit"* (both definitions agree in the case of coefficients in a field). Any polynomial may be decomposed into the product of an invertible constant by a product of irreducible polynomials. If the coefficients belong to a field or a unique factorization domain this decomposition is unique up to the order of the factors and the multiplication of any non-unit factor by a unit (and division of the unit factor by the same unit). When the coefficients belong to integers, rational numbers or a finite field, there are algorithms to test irreducibility and to compute the factorization into irreducible polynomials. These algorithms are not practicable for hand-written computation, but are available in any computer algebra system. Eisenstein's criterion can also be used in some cases to determine irreducibility.

Positional Notation

In modern positional numbers systems, such as the decimal system, the digits and their positions in the representation of an integer, for example, 45, are a shorthand notation for a polynomial in the radix or base, in this case, $4 \times 10^1 + 5 \times 10^0$. As another example, in radix 5, a string of digits such as 132 denotes the (decimal) number $1 \times 5^2 + 3 \times 5^1 + 2 \times 5^0 = 42$. This representation is unique. Let b be a positive integer greater than 1. Then every positive integer a can be expressed uniquely in the form,

$$a = r_m b^m + r_{m-1} b^{m-1} + \cdots + r_1 b + r_0,$$

where m is a nonnegative integer and the r's are integers such that:

$$0 < r_m < b \text{ and } 0 \leq r_i < b \text{ for } i = 0, 1, \ldots, m - 1.$$

Other Applications

Polynomials serve to approximate other functions, such as the use of splines.

Polynomials are frequently used to encode information about some other object. The characteristic polynomial of a matrix or linear operator contains information about the operator's eigenvalues. The minimal polynomial of an algebraic element records the simplest algebraic relation satisfied by that element. The chromatic polynomial of a graph counts the number of proper colourings of that graph.

The term "polynomial", as an adjective, can also be used for quantities or functions that can be written in polynomial form. For example, in computational complexity theory the phrase *polynomial time* means that the time it takes to complete an algorithm is bounded by a polynomial function of some variable, such as the size of the input.

In computer graphics they are used to interpolate between values to evaluate animation of dynamic graphical objects.

ZEROES OF POLYNOMIAL

A polynomial having value zero (0) is known as zero polynomial. Actually, the term 0 is itself zero polynomial. It is a constant polynomial whose all the coefficients are equal to 0. For a polynomial, there may be few (one or more) values of the variable for which the polynomial may result in zero. These values are known as zeros of a polynomial. We can say that the zeroes of a polynomial are defined as the points where the polynomial equals to zero on the whole.

If the coefficients of following the form of the polynomial: $a_n x^n + a_{n-1} x^{n-1} + a_{n-2} x^{n-2} + ... + a_2 x^2 + a_2 x^2 + a_1 x + a_0$ are zero, then it will become zero polynomial. i.e $a_n = a_{n-1} = a_{n-2} = ... = a_0 = 0$. Thus, the polynomial will become 0 and may be written as P(x)=0.

Zero Polynomial Function

The zero polynomial function is defined as the polynomial function with the value of zero. i.e. the function whose value is 0, is termed as a zero polynomial function. Zero polynomial does not have any nonzero term. It is represented as: P(x) = 0. Thus, we can say that a polynomial function which is equal to zero, is called zero polynomial function. It also is known as zero map. The graph of the zero polynomial is X axis.

Zero Quadratic Polynomial

The quadratic polynomial having all the coefficients equal to zero is known as zero quadratic polynomial. The general term of a quadratic polynomial is: $P(x) = ax^2 + bx + c$. If in above quadratic polynomial, the coefficients are zero; i.e. a = b = c = 0, then the polynomial is termed as a zero quadratic polynomial.

1. $0.x^2 + 0.x + 0$ is a zero quadractic polynomial whose values are zero.

2. Find the additive identities of the following polynomials: 1) x-3 and 2) $x^2 - 3x + 5$

Solution: 1. Additive identity = 0.x+0 and 2. Additive identity = $0.x^2 + 0.x + 0$.

Finding Zeroes of a Polynomial

1. The zero of a polynomial is the value of the which polynomial gives zero. Thus, in order to find zeros of the polynomial, we simply equate polynomial to zero and find the possible values of variables.

2. Let P(x) be a given polynomial. To find zeros, set this polynomial equal to zero. i.e. P(x) = 0. Now, this becomes a polynomial equation. Solve this equation and find all the possible values of variables by factorizing the polynomial equation.

3. These are the values of x which make polynomial equal to zero; hence are called zeros of polynomial P(x). A number z is said to be a zero of a polynomial P(x) if and only if P(z) = 0.

Real and Complex Zeroes of Polynomials

When the roots of a polynomial are in the form of the real number, they are known as real zeros whereas complex numbers are written as a $\pm$ ib, where a is called real part and b is known as the imaginary part. The complex zeros are found in pairs, such as a + ib and a − ib.

1. Find the zeroes of polynomial $6x^2 + 7x - 2$.

Solution: To find zeros, set the polynomial equal to zero P(x)=0 i.e. $6x^2 + 7x - 2 = 0$

$6x^2 + 4x - 3x - 2 = 0$ then, 2x(3x+2)-1(3x+2)=0

(3x+2)(2x-1)=0, $x = -\dfrac{2}{3}, \dfrac{1}{2}$.

2. Find the zeroes of polynomial $(x-3)^2 + 4$.

Solution: To find zeros, set the polynomial equal to zero P(x)=0 i.e. $(x-3)^2 + 4 = 0$

$(x-3)^2 + -4$ then, $x - 3 = \pm 2i$ and $x = 3 \pm 2i$

Thus, two zeros are 3 + 2i and 3 − 2i.

References

- Variable-mathematics, definition-1816: techopedia.com, Retrieved 29 March, 2019

- Bronstein, Manuel; et al., eds. (2006). Solving Polynomial Equations: Foundations, Algorithms, and Applications. Springer. ISBN 978-3-540-27357-8

- Zeroes-of-polynomial: mathstips.com, Retrieved 10 July, 2019

- Tabak, John (2014). Algebra: Sets, Symbols, and the Language of Thought. Infobase Publishing. p. 40. ISBN 978-0-8160-6875-3

- Algebraic-expressions, maths: byjus.com, Retrieved 30 January, 2019

- Foerster, Paul A. (2006). Algebra and Trigonometry: Functions and Applications, Teacher's Edition (Classics ed.). Upper Saddle River, NJ: Prentice Hall. ISBN 0-13-165711-9

- Algebraic-expressions, maths: byjus.com, Retrieved 16 May, 2019

Algebraic Functions and Equations

Algebraic functions can be classified into linear function, quadratic function, cubic function, quartic function, etc. Algebraic equations include linear equation, quadratic equation, cubic equation, quantic equation, etc. All these algebraic functions and equations have been carefully analyzed in this chapter.

ALGEBRAIC FUNCTION

A function $y = f(x_1,..,x_n)$ of the variables $x_1,..,x_n$ that satisfies an equation

$$F(y, x,..,x) \quad 0,$$

where F is an irreducible polynomial in $y, x_1,..,x_n$ with coefficients in some field K, known as the field of constants. The algebraic function is said to be defined over this field, and is called an algebraic function over the field K. The polynomial $F(y, x_1,..,x_n)$ is often written in powers of the variable y, so that equation $F(y, x_1,..,x_n) = 0,$ assumes the form

$$P_k(x_1,..,x_n)\, y^k + P_{k-1}(x_1,..,x_n) y^{k-1} + ... +$$
$$+ P_0(x_1,..,x_n) = 0,$$

where $P_k(x_1,..,x_n),.., P_0(x_1,..,x_n)$ are polynomials in $x_1,..,x_n$, and with $P_k(x_1,..,x_n) \neq 0$. The number k is the degree of F with respect to y, and is called the degree of the algebraic function. If $k = 1$, an algebraic function may be represented as a quotient

$$y = -\frac{P_0(x_1,..,x_n)}{P_1(x_1,..,x_n)}$$

of polynomials, and is called a rational function of $x_1,..,x_n$. For $k=2,3,4$, an algebraic function can be expressed as square and cube roots of rational functions in the variables $x_1,..,x_n$ if $k > 4$ this is impossible in general.

The theory of algebraic functions was studied in the past from three different points of view: the function-theoretical point of view taken, in particular, by N.H. Abel, K. Weierstrass and B. Riemann; the arithmetic-algebraic point of view taken by R. Dedekind, H. Weber and K. Hensel; and the algebraic-geometrical point of view, which originated with the studies of A. Clebsch, M. Noether and others. The first direction of the theory of algebraic functions of a single variable is connected with the study of algebraic functions over the field of complex numbers, in which they are regarded as meromorphic functions on Riemann surfaces and complex manifolds; the most important methods applied are the geometrical and topological methods of the theory of analytic functions. The arithmetic-algebraic approach involves the study of algebraic functions over arbitrary fields. The methods employed are purely algebraic. The theory of valuations and extensions of fields are especially important. In the algebraic-geometrical approach algebraic functions are considered to be rational functions on an algebraic variety, and are studied by methods of algebraic geometry. These three points of view originally differed not only in their methods and their ways of reasoning, but also in their terminology. This differentiation has by now become largely arbitrary, since function-theoretical studies involve the extensive use of algebraic methods, while many results obtained at first using function-theoretical and topological methods can be successfully applied to more general fields using algebraic analogues of these methods.

Algebraic Functions of One Variable

Over the field C of complex numbers, an algebraic function of one variable $y = f(x)$ or $y(x)$ for short) is a k-valued analytic function. If D(x) is the discriminant of the polynomial,

$$F(x,y) = P_k(x)y^k + ... + P_1(x)y + P_0(x),$$
$$P_k(x) \neq 0,$$

(i.e. of the polynomial for which $F(x, f(x)) = 0$), which is obtained by eliminating y from the equations,

$$F(x.y) = 0, \quad \frac{\partial F(x,y)}{\partial y} = 0$$

to yield the equation,

$$P_k(x)D(x) = 0,$$

then the roots x ,.., x of this last equation are known as the critical values of $y = f(x)$. The complementary set $G = C \setminus \{x_1, .., x_m\}$ is known as the non-critical set. For any point $x_0 \in G$ equation $F(x,y) = P_k(x)y^k + ... + P_1(x)y + P_0(x)$, has k different roots $y_0^1, .., y_0^k$ and the condition,

$$\partial F \frac{(x_0, y_0^j)}{\partial y} \neq 0, \quad j = 1, .., k,$$

is satisfied. According to the implicit function theorem, in a neighbourhood of the point x_0 there exist k single-valued analytic functions $f_0^1(x), .., f_0^k(x)$ which satisfy the conditions,

$$f_0^j(x_0) = y_0^j \qquad F(x, f_0^j(x)) = 0$$

and which can be decomposed into a convergent series,

$$f_0^j(x) = y_0^j + \alpha_1^j(x - x_0) + \alpha_2^j(x - x_0)^2 + ...$$

Thus, for each point $x_0 \in G$ one can construct k elements of an analytic function, known as the function elements with centre at the point x_0. For any two points $x_1, x_2 \in G$, any elements $f_1^i(x)$ and $f_2^i(x)$ with centres at x_1 and x_2, respectively, are derived from each other by analytic continuation along some curve in G; in particular, any two elements with the same centres are also connected in this way. If x_0 is a critical point of an algebraic function, then two cases are possible: 1) x_0 is a root of the discriminant, i.e. $D(x_0)=0$, but $P_k(x_0) \neq 0$; or 2) $P_k(x_0)=0$.

Case: Let K_0 be a small disc with centre at x_0 which does not contain other critical points, and let $f_1'(x),.., f_k'(x)$ be a system of regular elements with centre at $x' \in K_0$, $x' \neq x_0$. These functions remain bounded as $x \to x_0$. Furthermore, let D be the circle with centre x_0 passing through x'; it is completely contained inside K_o. The analytic continuation of some given element, e.g. $f_1'(x)$, along D (in, say, the clockwise direction), yields an element $f'(x)$ which also belongs to the system of elements with centre x'. This system consists of k elements, and a minimum required finite number $\alpha_1 \leq k$ of such turns yield the initial element $f_1'(x)$. One obtains a subsystem $f_1'(x),.., f_{\alpha_1}'(x)$ of elements with centre x'; each one of these elements may be obtained by analytic continuation of the other by a number of turns around the point x_0; such a subsystem is known as a cycle. Any system $f_1(x),.., f_k(x)$ can be decomposed into a number of non-intersecting cycles,

$$\{f_1'(x),.., f_{\alpha_1}'(x)\}$$
$$\{f_{\alpha_1}'+1(x),.., f_{\alpha_1+...\alpha_2}'(x)\},...,$$
$$\{f_{\alpha_1}'+...\alpha_{s-1}+1(x),.., f_{\alpha_1+...+\alpha_s}'(x)\},$$

$\alpha_1+...\alpha_s = k$. If $\alpha_1 > 1$, then the element $f_1'(x)$ is not a single-valued function of x in the disc K_0, but is a single-valued analytic function of the parameter $\tau = (x-x_0)^{1/\alpha_1}$ in a neighbourhood of $\tau = 0$. In a certain neighbourhood of this point the elements $f_1'(x),..., f_{\alpha_1}'(x)$ of the first cycle can be represented as convergent series,

$$f_1'(x)=\sum_{i=0}^{\infty} \alpha_i^1 \tau^i = \sum_{i=0}^{\infty} \alpha_i^1 (x-x_0)^{i/\alpha_1},$$

$$f_{\alpha_1}'(x)=\sum_{i=0}^{\infty} \alpha_i^{\alpha_1} \tau^i = \sum_{i=0}^{\infty} \alpha_i^{\alpha_1} (x-x_0)^{i/\alpha_1};$$

and similar expansions also take place for the elements of other cycles. Such expansions of elements by fractional degrees of the difference $x - x_0$, where x_0 is a critical point, are known as Puiseux series. The transformation $\tau \to \tau r$, $_r = e^{2\pi i/\alpha}$; which corresponds to one turn around x_0, converts the Puiseux series of elements in one cycle into each other in cyclic order, i.e. there is a cyclic permutation of the series and of the corresponding elements. To turns around the critical point correspond permutations of the elements with centre at this point; these permutations consist of cycles of the orders, $\alpha_{1,..}, \alpha_s \sum \alpha_i = k$. The permutations defined in this way constitute the monodromy group of the algebraic function. If at least one element α_i is greater than 1, the critical point x_0 is called an algebraic branch point of the algebraic function; the numbers α_i (sometimes $\alpha_1 - 1$) are called the branch indices (or branch orders) of the algebraic function.

Case: If $P_k(x)y$ is substituted for Y one returns to case above; one obtains expansions similar to $f_1'(x) = \sum_{i=0}^{\infty} \alpha_i^1 \dots (x-x_0)^{i/\alpha_1}$;;which may contain a finite number of terms with negative indices:

$$f(x) = \sum_{i=-p}^{\infty} \alpha_i \ \tau^i = \sum_{i=-p}^{\infty} \alpha_i \ (x-x_0)^{i/\alpha}.$$

If $p > 0$, the point X_0 is a pole of order P of the algebraic function. An algebraic function is usually considered on the Riemann sphere S, i.e. on the complex plane completed by the point at infinity $x = \infty$. The introduction of the variable $\tau = 1/x$ reduces this case to the previous case; in a neighbourhood of the point $\tau = 0 (x = \infty)$ one has the expansion:

$$y(x) = \sum_{j=-r}^{\infty} \alpha j \ \tau^{j/\alpha} = \sum_{j=-r}^{\infty} \alpha j \ x^{-j/\alpha}.$$

If $r > 0$, then the point $x = \infty$ is called a pole of order r.

The parameter of the expansion in the series $f_0^j(x) = y_0^j + \alpha_1^j(x-x_0) + \alpha_2^j(x-x_0)^2 + \dots$, $f_1'(x) = \sum_{i=0}^{\infty} \alpha_i^1 \dots (x-x_0)^{i/\alpha_1}$;; $f(x) = \sum_{i=-p}^{\infty} \alpha_i \ \tau^i = \sum_{i=-p}^{\infty} \alpha_i \ (x-x_0)^{i/\alpha}$, $y(x) = \sum_{j=-r}^{\infty} \alpha j \ \tau^{j/\alpha} = \sum_{j=-r}^{\infty} \alpha j \ x^{-j/\alpha}$ is called the local uniformizing parameter for the algebraic function. If x_0 is a non-critical point of the algebraic function, then $\tau = x - x_0$ can be taken as parameter; if, on the other hand, is a critical point, the root $(x-x_0)^{1/\alpha}$ (where α is a positive integer) can be taken as such a parameter. The population of all elements of an algebraic function described above forms the complete algebraic function in the sense of Weierstrass. Algebraic functions have no singularities other than algebraic branch points and poles. The converse proposition is also true: A function $y = f(x)$ which is analytic, is not more than s-valued at all points of the Riemann sphere except for a finite number of points $x_1, .., x_m$ and $x = \infty$, and has at such points only poles or algebraic branch points, is an algebraic function of degree $k \leq s$.

The Riemann surface of a complete algebraic function is compact and is a k-sheeted covering of the Riemann sphere, branch points can be the critical points and the point $x = \infty$. Algebraic functions are the only class of functions with a compact Riemann surface. The genus of the Riemann surface of an algebraic function is important; it is called the genus of the algebraic function. It can be calculated by the Riemann–Hurwitz formula. The genus of a rational function is zero, and its Riemann surface is the Riemann sphere. The Riemann surface of an elliptic function that satisfies a third- or fourth-degree equation is a torus; the genus of such a function is one.

The universal covering Riemann surface of an algebraic function is a simply-connected two-dimensional manifold, i.e. it has a trivial fundamental group and is conformally equivalent either to the Riemann sphere, the complex plane or the interior of the unit disc. In the first case the algebraic function is a rational, in the second case it is an elliptic, while in the third case it is a general function.

The uniformization problem of algebraic functions is closely connected with the theory of Riemann surfaces of algebraic functions. The function $y = f(x)$ can be uniformized if y and x are representable as single-valued analytic functions

$$y = y(t), \quad x = x(t)$$

of a parameter t, which identically satisfy equation $F(x,y) = P_k(x)y^k + ... + P_1(x)y + P_0(x)$. The uniformization problem is locally solved by a local uniformizing parameter; however, it is the solution "in the large" that is of interest. If $k = 1$, i.e. if $y(x)$ is a rational function of x, this parameter may be the variable $x - x_0$; if $k = 2$, then uniformization is attained with the aid of a rational or a trigonometric function. For instance, if $y(x)$ satisfies the equation,

$$y^2 - x^2 = 1$$

one can take,

$$y = \frac{t^2 + 1}{t^2 - 1}, \quad x = \frac{2t}{t^2 - 1}$$

or

$$y = \sec t, \quad x = \operatorname{tg} t.$$

If $k = 3, 4$ in the case of an algebraic function of genus one, uniformization is achieved using elliptic functions. Finally, if $k > 4$ and the genus of the algebraic function is higher than one, uniformization is realized using automorphic functions.

Algebraic Functions of Several Variables

If f is an algebraic function in the variables $x_1, ..., x_n$, then the set of all rational functions $R(y, x_1, ..., x_n)$ forms a field K_f, coinciding with the field of rational functions on the algebraic hypersurface in $(n+1)$-dimensional space defined by the equation $F(y, x_1, ..., x_n) = 0$. If the field of constants k is the field of complex numbers C and if $n = 1$, then K_f is identical with the field of meromorphic functions on the Riemann surface of the algebraic function. The field K_f is an extension of finite type of the field of constants k of transcendence degree n (cf. Extension of a field). In particular, any $n+1$ elements of this field are connected by an algebraic equation, so that each of them defines an algebraic function of the remaining elements. Any extension K of finite type of a field k of transcendence degree n is known as an algebraic function field in n variables (or, sometimes, as a function field). Each such field contains a purely transcendence extension $k(x_1, ..., x_n)$ of the field k (called the field of rational functions in n variables). Any element $y \in K$ satisfies some algebraic equation $\Phi(y, x_1, ..., x_n) = 0$, and can be considered as an algebraic function in the variables $x_1, ..., x_n$. Each field K of algebraic functions in n variables is isomorphic to the field of rational functions on some algebraic variety of dimension n, which is called a model of K. If the field of constants k is algebraically closed and of characteristic zero, then each algebraic function field has a non-singular projective model. Let S be the set of all non-trivial valuations of an algebraic function field K which are non-negative on the field of constants. If provided with the natural topology, it is known as the abstract Riemann surface of the field K. In the case of algebraic functions in one variable, the Riemann surface coincides with the set of non-singular projective models, which in this case is uniquely defined up to an isomorphism. Many concepts and results in algebraic geometry on the model of a field K can be restated in the language of the theory of valuations of fields. A particularly close analogy holds for algebraic functions in one variable, the theory of which is practically identical with the theory of algebraic curves.

Each algebraic function field in one variable is the field of fractions of a Dedekind ring, so that many results and concepts of the theory of divisibility in algebraic number fields can be applied to function fields. Many problems and constructions in algebraic number theory motivate similar problems and constructions in fields of algebraic functions and vice versa. For instance, the application of a Puiseux expansion to the theory of algebraic numbers led to the genesis of the P-adic method in number theory, due to Hensel. Class field theory, which had originally belonged to the domain of algebraic numbers, was subsequently applied to functions. An especially close analogy exists between algebraic number fields and algebraic function fields over a finite field of constants. For instance, the concept of a zeta-function is defined for the latter and the analogue of the Riemann hypothesis has been demonstrated for algebraic function fields.

LINEAR FUNCTION

In mathematics, the term linear function refers to two distinct but related notions:

- In calculus and related areas, a linear function is a function whose graph is a straight line, that is a polynomial function of degree one or zero.

- In linear algebra, mathematical analysis, and functional analysis, a linear function is a linear map. In this case, and in case of possible ambiguity, the name affine function is often used for the concept above.

As a Polynomial Function

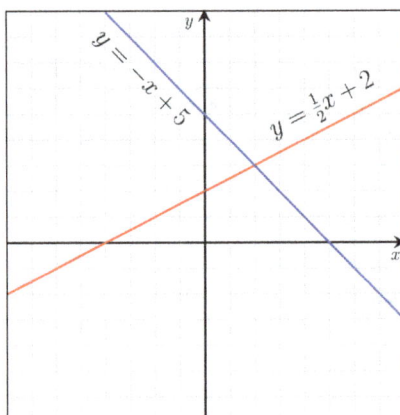

Graphs of two linear (polynomial) functions.

In calculus, analytic geometry and related areas, a linear function is a polynomial of degree one or less, including the zero polynomial (the latter not being considered to have degree zero).

When the function is of only one variable, it is of the form

$$f(x) = ax + b,$$

where a and b are constants, often real numbers. The graph of such a function of one variable is a nonvertical line. a is frequently referred to as the slope of the line, and b as the intercept.

For a function $f(x_1,\ldots,x_k)$ of any finite number of independent variables, the general formula is

$$f(x_1,\ldots,x_k) = b + a_1 x_1 + \ldots + a_k x_k,$$

and the graph is a hyperplane of dimension k.

A constant function is also considered linear in this context, as it is a polynomial of degree zero or is the zero polynomial. Its graph, when there is only one independent variable, is a horizontal line.

In this context, the other meaning (a linear map) may be referred to as a homogeneous linear function or a linear form. In the context of linear algebra, this meaning (polynomial functions of degree 0 or 1) is a special kind of affine map.

As a Linear Map

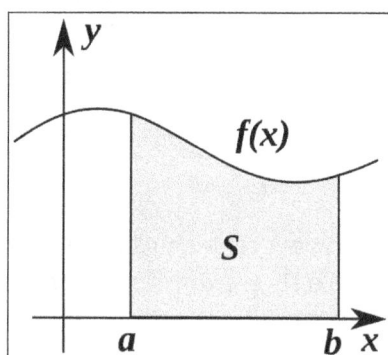

The integral of a function is a linear map from the
vector space of integrable functions to the real numbers.

In linear algebra, a linear function is a map f between two vector spaces that preserves vector addition and scalar multiplication:

$$f(\mathrm{x} + \mathrm{y}) = f(\mathrm{x}) + f(\mathrm{y})$$
$$f(a\mathrm{x}) = af(\mathrm{x}).$$

Here a denotes a constant belonging to some field K of scalars (for example, the real numbers) and x and y are elements of a vector space, which might be K itself.

Some authors use "linear function" only for linear maps that take values in the scalar field; these are also called linear functionals.

The "linear functions" of calculus qualify as "linear maps" when (and only when) $b = 0$., or, equivalently, when the constant $f([0,\ldots,0]) = 0$,. Geometrically, the graph of the function must pass through the origin.

QUADRATIC FUNCTION

In algebra, a quadratic function, a quadratic polynomial, a polynomial of degree 2, or simply a quadratic, is a polynomial function with one or more variables in which the highest-degree term is of the second degree. For example, a quadratic function in three variables x, y, and z contains exclusively terms x^2, y^2, z^2, xy, xz, yz, x, y, z, and a constant:

$$f(x, y, z) = ax^2 + by^2 + cz^2 + dxy + exz + fyz + gx + hy + iz + j,$$

with at least one of the coefficients a, b, c, d, e, or f of the second-degree terms being non-zero.

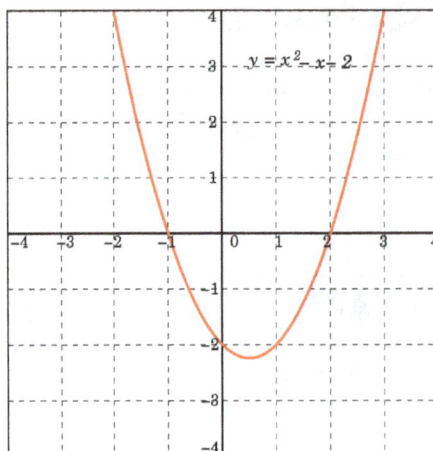

A quadratic polynomial with two real roots (crossings of the x axis) and hence no complex roots. Some other quadratic polynomials have their minimum above the x axis, in which case there are no real roots and two complex roots.

A *univariate* (single-variable) quadratic function has the form,

$$f(x) = ax^2 + bx + c, \quad a \neq 0$$

in the single variable x. The graph of a univariate quadratic function is a parabola whose axis of symmetry is parallel to the y-axis.

If the quadratic function is set equal to zero, then the result is a quadratic equation. The solutions to the univariate equation are called the roots of the univariate function.

The bivariate case in terms of variables x and y has the form,

$$f(x, y) = ax^2 + by^2 + cxy + dx + ey + f$$

with at least one of a, b, c not equal to zero, and an equation setting this function equal to zero gives rise to a conic section (a circle or other ellipse, a parabola, or a hyperbola).

In general there can be an arbitrarily large number of variables, in which case the resulting surface is called a quadric, but the highest degree term must be of degree 2, such as x^2, xy, yz, etc.

Terminology

Coefficients

The coefficients of a polynomial are often taken to be real or complex numbers, but in fact, a polynomial may be defined over any ring.

Degree

When using the term "quadratic polynomial", authors sometimes mean "having degree exactly 2", and sometimes "having degree at most 2". If the degree is less than 2, this may be called a "degenerate case". Usually the context will establish which of the two is meant.

Sometimes the word "order" is used with the meaning of "degree", e.g. a second-order polynomial.

Variables

A quadratic polynomial may involve a single variable x (the univariate case), or multiple variables such as x, y, and z (the multivariate case).

The One-variable Case

Any single-variable quadratic polynomial may be written as:

$$ax^2 + bx + c,$$

where x is the variable, and a, b, and c represent the coefficients. In elementary algebra, such polynomials often arise in the form of a quadratic equation $ax^2 + bx + c = 0$. The solutions to this equation are called the roots of the quadratic polynomial, and may be found through factorization, completing the square, graphing, Newton's method, or through the use of the quadratic formula. Each quadratic polynomial has an associated quadratic function, whose graph is a parabola.

Bivariate Case

Any quadratic polynomial with two variables may be written as:

$$f(x, y) = ax^2 + by^2 + cxy + dx + ey + f,$$

where x and y are the variables and a, b, c, d, e, and f are the coefficients. Such polynomials are fundamental to the study of conic sections, which are characterized by equating the expression for $f(x, y)$ to zero. Similarly, quadratic polynomials with three or more variables correspond to quadric surfaces and hypersurfaces. In linear algebra, quadratic polynomials can be generalized to the notion of a quadratic form on a vector space.

Forms of a Univariate Quadratic Function

A univariate quadratic function can be expressed in three formats:

- $f(x) = ax^2 + bx + c$ is called the standard form,
- $f(x) = a(x - r_1)(x - r_2)$ is called the factored form, where r_1 and r_2 are the roots of the quadratic function and the solutions of the corresponding quadratic equation.
- $f(x) = a(x - h)^2 + k$ is called the vertex form, where h and k are the x and y coordinates of the vertex, respectively.

The coefficient a is the same value in all three forms. To convert the standard form to factored form, one needs only the quadratic formula to determine the two roots r_1 and r_2. To convert the standard form to vertex form, one needs a process called completing the square. To convert the factored form (or vertex form) to standard form, one needs to multiply, expand and/or distribute the factors.

Graph of the Univariate Function

Regardless of the format, the graph of a univariate quadratic function $f(x) = ax^2 + bx + c$ is a parabola. Equivalently, this is the graph of the bivariate quadratic equation $y = ax^2 + bx + c$.

- If $a > 0$, the parabola opens upwards.

- If $a < 0$, the parabola opens downwards.

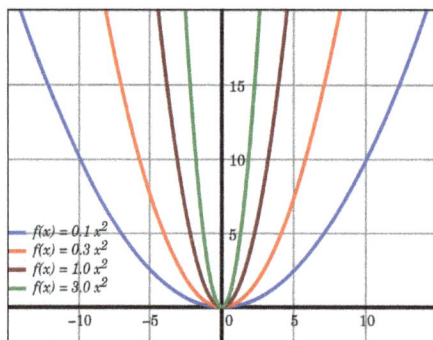

$$f(x) = ax^2 \big|_{a=\{0.1, 0.3, 1, 3\}}$$

The coefficient a controls the degree of curvature of the graph; a larger magnitude of a gives the graph a more closed (sharply curved) appearance.

The coefficients b and a together control the location of the axis of symmetry of the parabola (also the x-coordinate of the vertex) which is at

$$x = -\frac{b}{2a}.$$

The coefficient c controls the height of the parabola; more specifically, it is the height of the parabola where it intercepts the y-axis.

Vertex

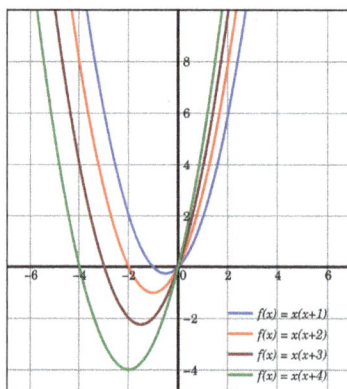

$$f(x) = x^2 + bx \big|_{b=\{1,2,3,4\}}$$

The vertex of a parabola is the place where it turns; hence, it is also called the turning point. If the quadratic function is in vertex form, the vertex is (h, k). Using the method of completing the square, one can turn the standard form

$$f(x) = ax^2 + bx + c$$

into

$$\begin{aligned} f(x) &= ax^2 + bx + c \\ &= a(x-h)^2 + k \\ &= a\left(x - \frac{-b}{2a}\right)^2 + \left(c - \frac{b^2}{4a}\right), \end{aligned}$$

so the vertex, (h, k), of the parabola in standard form is

$$\left(-\frac{b}{2a}, c - \frac{b^2}{4a}\right).$$

If the quadratic function is in factored form

$$f(x) = a(x - r_1)(x - r_2)$$

the average of the two roots, i.e.,

$$\frac{r_1 + r_2}{2}$$

is the x-coordinate of the vertex, and hence the vertex (h, k) is

$$\left(\frac{r_1 + r_2}{2}, f\left(\frac{r_1 + r_2}{2}\right)\right).$$

The vertex is also the maximum point if $a < 0$, or the minimum point if $a > 0$.

The vertical line

$$x = h = -\frac{b}{2a}$$

that passes through the vertex is also the axis of symmetry of the parabola.

Maximum and Minimum Points

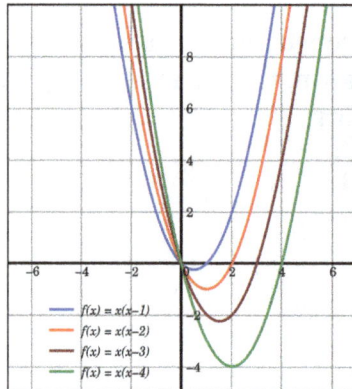

$$f(x) = x^2 + bx\,\big|_{b=\{-1,-2,-3,-4\}}$$

Using calculus, the vertex point, being a maximum or minimum of the function, can be obtained by finding the roots of the derivative:

$$f(x) = ax^2 + bx + c \quad \Rightarrow \quad f'(x) = 2ax + b.$$

x is a root of $f'(x)$ if $f'(x) = 0$ resulting in

$$x = -\frac{b}{2a}$$

with the corresponding function value

$$f(x) = a\left(-\frac{b}{2a}\right)^2 + b\left(-\frac{b}{2a}\right) + c = c - \frac{b^2}{4a},$$

so again the vertex point coordinates, (h, k), can be expressed as

$$\left(-\frac{b}{2a}, c - \frac{b^2}{4a}\right).$$

Roots of the Univariate Function

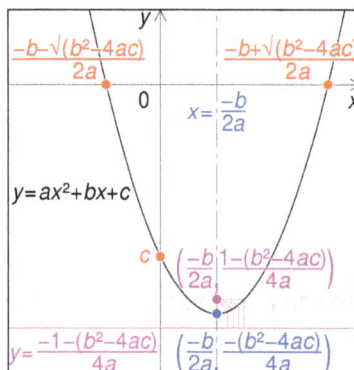

Graph of $y = ax^2 + bx + c$, where a and the discriminant $b^2 - 4ac$ are positive, with

- Roots and y-intercept in red.
- Vertex and axis of symmetry in blue.
- Focus and directrix in pink.

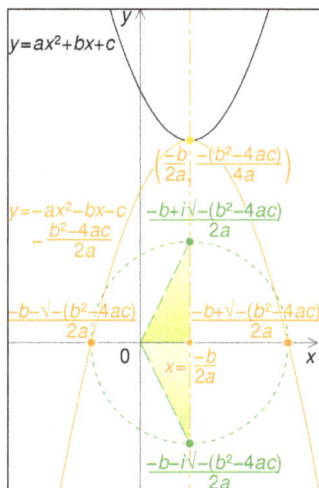

Visualisation of the complex roots of $y = ax^2 + bx + c$: the parabola is rotated 180° about its vertex (orange). Its x-intercepts are rotated 90° around their mid-point, and the Cartesian plane is interpreted as the complex plane (green).

Exact Roots

The roots (or *zeros*), r_1 and r_2, of the univariate quadratic function

$$f(x) = ax^2 + bx + c$$
$$= a(x - r_1)(x - r_2),$$

are the values of x for which $f(x) = 0$.

When the coefficients a, b, and c, are real or complex, the roots are

$$r_1 = \frac{-b - \sqrt{b^2 - 4ac}}{2a},$$

$$r_2 = \frac{-b + \sqrt{b^2 - 4ac}}{2a}.$$

Upper Bound on the Magnitude of the Roots

The modulus of the roots of a quadratic $ax^2 + bx + c$ can be no greater than $\dfrac{\max(|a|, |b|, |c|)}{|a|} \times \phi$,

where ϕ is the golden ratio $\dfrac{1 + \sqrt{5}}{2}$.

The Square Root of a Univariate Quadratic Function

The square root of a univariate quadratic function gives rise to one of the four conic sections, almost always either to an ellipse or to a hyperbola.

If $a > 0$ then the equation $y = \pm\sqrt{ax^2 + bx + c}$ describes a hyperbola, as can be seen by squaring both sides. The directions of the axes of the hyperbola are determined by the ordinate of the minimum point of the corresponding parabola $y_p = ax^2 + bx + c$. If the ordinate is negative, then the hyperbola's major axis (through its vertices) is horizontal, while if the ordinate is positive then the hyperbola's major axis is vertical.

If $a < 0$ then the equation $y = \pm\sqrt{ax^2 + bx + c}$ describes either a circle or other ellipse or nothing at all. If the ordinate of the maximum point of the corresponding parabola $y_p = ax^2 + bx + c$ is positive, then its square root describes an ellipse, but if the ordinate is negative then it describes an empty locus of points.

Iteration

To iterate a function $f(x) = ax^2 + bx + c$, one applies the function repeatedly, using the output from one iteration as the input to the next.

One cannot always deduce the analytic form of $f^{(n)}(x)$, which means the nth iteration of $f(x)$. (The superscript can be extended to negative numbers, referring to the iteration of the inverse of $f(x)$ if the inverse exists). But there are some analytically tractable cases.

For example, for the iterative equation,

$$f(x) = a(x - c)^2 + c$$

one has,

$$f(x) = a(x - c)^2 + c = h^{(-1)}(g(h(x))),$$

where,

$$g(x) = ax^2 \text{ and } h(x) = x - c.$$

So by induction,

$$f^{(n)}(x) = h^{(-1)}(g^{(n)}(h(x)))$$

can be obtained, where $g^{(n)}(x)$ can be easily computed as:

$$g^{(n)}(x) = a^{2^n - 1} x^{2^n}.$$

Finally, we have:

$$f^{(n)}(x) = a^{2^n - 1}(x - c)^{2^n} + c$$

as the solution.

The logistic map,

$$x_{n+1} = rx_n(1-x_n), \quad 0 \le x_0 < 1$$

with parameter 2<r<4 can be solved in certain cases, one of which is chaotic and one of which is not. In the chaotic case r=4 the solution is:

$$x_n = \sin^2(2^n \theta \pi)$$

where the initial condition parameter is given by $\theta = \dfrac{1}{\pi} \sin^{-1}(x_0^{1/2})$. For rational θ, after a finite number of iterations x_n maps into a periodic sequence. But almost all θ are irrational, and, for irrational θ, x_n never repeats itself – it is non-periodic and exhibits sensitive dependence on initial conditions, so it is said to be chaotic.

The solution of the logistic map when r=2 is:

$$x_n = \frac{1}{2} - \frac{1}{2}(1-2x_0)^{2^n}$$

for $x_0 \in [0,1)$. Since $(1-2x_0) \in (-1,1)$ for any value of x_0 other than the unstable fixed point 0, the term $(1-2x_0)^{2^n}$ goes to 0 as n goes to infinity, so x_n goes to the stable fixed point $\frac{1}{2}$.

Bivariate (Two Variable) Quadratic Function

A bivariate quadratic function is a second-degree polynomial of the form:

$$f(x,y) = Ax^2 + By^2 + Cx + Dy + Exy + F$$

where A, B, C, D, and E are fixed coefficients and F is the constant term. Such a function describes a quadratic surface. Setting $f(x,y)$ equal to zero describes the intersection of the surface with the plane $z = 0$, which is a locus of points equivalent to a conic section.

Minimum/Maximum

If $4AB - E^2 < 0$ the function has no maximum or minimum; its graph forms an hyperbolic paraboloid.

If $4AB - E^2 > 0$ the function has a minimum if A>0, and a maximum if A<0; its graph forms an elliptic paraboloid. In this case the minimum or maximum occurs at (x_m, y_m) where:

$$x_m = -\frac{2BC - DE}{4AB - E^2},$$

$$y_m = -\frac{2AD - CE}{4AB - E^2}.$$

If $4AB - E^2 = 0$ and $DE - 2CB = 2AD - CE \ne 0$ the function has no maximum or minimum; its graph forms a parabolic cylinder.

If $4AB - E^2 = 0$ and $DE - 2CB = 2AD - CE = 0$ the function achieves the maximum/minimum at a line—a minimum if A>0 and a maximum if A<0; its graph forms a parabolic cylinder.

CUBIC FUNCTION

A cubic function has the standard form of $f(x) = ax^3 + bx^2 + cx + d$. The "basic" cubic function is $f(x) = x^3$. You can see it in the graph below. In a cubic function, the highest power over the x variable(s) is 3.

Critical and Inflection Points

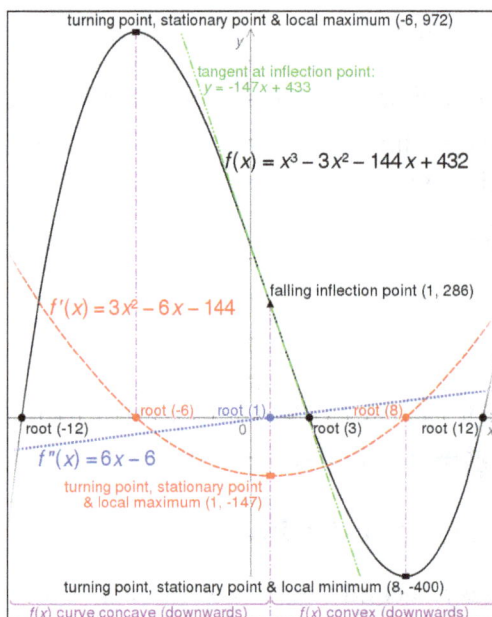

turning point, stationary point & local maximum (-6, 972)

tangent at inflection point: $y = -147x + 433$

$f(x) = x^3 - 3x^2 - 144x + 432$

falling inflection point (1, 286)

$f'(x) = 3x^2 - 6x - 144$

root (-6)   root (1)   root (8)

root (-12)   root (3)   root (12)

$f''(x) = 6x - 6$

turning point, stationary point & local maximum (1, -147)

turning point, stationary point & local minimum (8, -400)

f(x) curve concave (downwards)   f(x) convex (downwards)

The roots, turning points, stationary points, inflection point and concavity of a cubic polynomial $x^3 - 3x^2 - 144x + 432$ (black line) and its first and second derivatives (red and blue).

The critical points of a cubic function are its stationary points, that is the points where the slope of the function is zero. Thus the critical points of a cubic function f defined by

$$f(x) = ax^3 + bx^2 + cx + d,$$

occur at values of x such that the derivative

$$3ax^2 + 2bx + c = 0$$

of the cubic function is zero.

The solutions of that equation are the critical points of the cubic equation and are given, using the quadratic formula, by

$$x_{\text{critical}} = \frac{-b \pm \sqrt{b^2 - 3ac}}{3a}.$$

The expression inside the square root,

$$\Delta_0 = b^2 - 3ac,$$

determines what type of critical points the function has. If $\Delta_0 > 0$, then the cubic function has a local maximum and a local minimum. If $\Delta_0 = 0$, then the cubic's inflection point is the only critical point. If $\Delta_0 < 0$, then there are no critical points. In cases where $\Delta_0 \leq 0$, the cubic function is strictly monotonic. The adjacent diagram is an example of the case where $\Delta_0 > 0$. The other two cases do not have the local maximum or the local minimum but still have an inflection point.

The value of Δ_0 also plays an important role in determining the nature of the roots of the cubic equation and in the calculation of those roots.

The inflection point of a function is where that function changes concavity. The inflection point of our cubic function occurs at:

$$x_{\text{inflection}} = -\frac{b}{3a},$$

a value that is also important in solving the cubic equation. The cubic function has point symmetry about its inflection point.

All of the above assumes that the coefficients are real as well as the variable x.

QUARTIC FUNCTION

In mathematics, a quartic function, is a function of the form where a is nonzero, which is defined by a polynomial of degree four, called quartic polynomial. Sometimes the term biquadratic is used instead of quartic, but, usually, biquadratic function refers to a quadratic function of a square, having the form A quartic equation, or equation of the fourth degree, is an equation consisting in equating to zero a quartic polynomial, of the form where a ≠ 0. The derivative of a quartic function is a cubic function. Since a quartic function is defined by a polynomial of even degree, it has the same infinite limit when the argument goes to positive or negative infinity. If a is positive, then the function increases to positive infinity at both sides; and thus the function has a global minimum. Likewise, if a is negative, it decreases to negative infinity and has a global maximum. In both cases it may have, but not always, another local maximum and another local minimum. The degree four is the highest degree such that every polynomial equation can be solved by radicals.

Quartic function, the fourth degree polynomial $f(x) a_4 x^4 + a_3 x^3 + a_2 x^2 + a_1 x + a_0$

Transformation of the quartic polynomial from the general to the source form.

To get the source quartic function we plug the coordinates of translations:

$$x_0 = \frac{a_{n-1}}{n \cdot a_n} = -\frac{a_3}{4a_4} \text{ and } y_0 = f(x_0) = \frac{-3a_3^4 + 16 a_4 a_3^2 a_2 - 64 a_4^2 a_3 a_1 + 256 a_4^3 a_0}{256 a_4^3}$$

(with changed signs) into general form of the quartic polynomial,

$$y + y_0 = a_4 (x + x_0)^4 + a_3 (x + x_0)^3 + a_2 (x + x_0)^2 + a_1 (x + x_0) + a_0,$$

after expanding and reducing obtained is the source quartic function

$$y = \alpha_4 x^4 + \alpha_2 x^2 \, \alpha_1 x \text{ where, } \alpha_2 = \frac{-3\alpha_3^2 + 8\alpha_4\alpha_2}{8\alpha_4} \text{ and } \alpha_1 \frac{\alpha_3^3 - 4\alpha_4\alpha_3\alpha_2 + 8\alpha_4^2\alpha_1}{8\alpha_4^2}.$$

The basic classification criteria applied to the source quartic polynomial shows the diagram.

$$y = \alpha_4 x^4 + \alpha_2 x^2 + \alpha_1 x$$

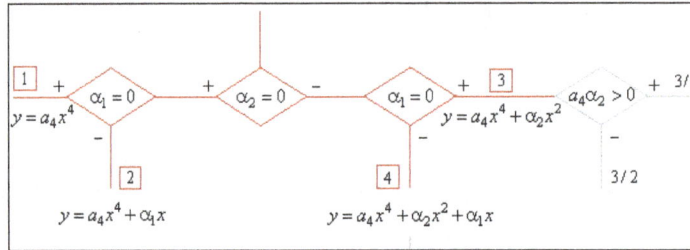

Thus, there are ten types (different shapes of graphs) of quartic functions. Applying additional criteria defined are the conditions remaining six types of the quartic polynomial functions to appear.

Observe that the basic criteria of the classification separates even and odd n^{th} degree polynomials called the power functions or monomials as the first type, since all coefficients a of the source function vanish.

Therefore, the first type of the qurtic polynomial.

$$y = \alpha_4 x^4 + \alpha_3 x^3 + \alpha_2 x^2 + \alpha_1 x + \alpha_0 \text{ or } y - y_0 = \alpha_4 (x - x_0)^4, \alpha_2 = 0 \text{ and } \alpha_1 = 0.$$

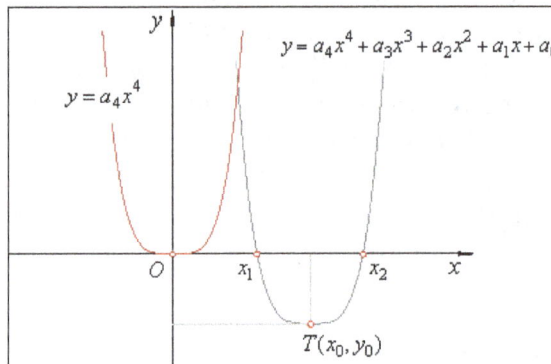

If $\alpha_4 \cdot y_0 \leq 0$ then the roots are $x_{1,2} = x_0 \pm \sqrt[4]{-\dfrac{y_0}{\alpha_4}}$. The turning point $T(x_o, y_o)$.

QUINTIC FUNCTION

In algebra, a quintic function is a function of the form

$$g(x) = ax^5 + bx^4 + cx^3 + dx^2 + ex + f,$$

where a, b, c, d, e and f are members of a field, typically the rational numbers, the real numbers or the complex numbers, and a is nonzero. In other words, a quintic function is defined by a polynomial of degree five.

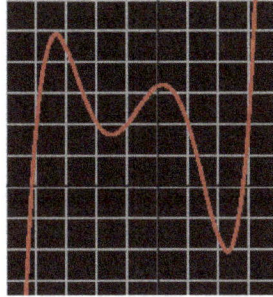

Graph of a polynomial of degree 5,
with 3 real zeros (roots) and 4 critical points.

If a is zero but one of the coefficients b, c, d, or e is non-zero, the function is classified as either a quartic function, cubic function, quadratic function or linear function.

Because they have an odd degree, normal quintic functions appear similar to normal cubic functions when graphed, except they may possess an additional local maximum and local minimum each. The derivative of a quintic function is a quartic function.

Setting $g(x) = 0$ and assuming $a \neq 0$ produces a quintic equation of the form:

$$ax^5 + bx^4 + cx^3 + dx^2 + ex + f = 0.$$

Solving quintic equations in terms of radicals was a major problem in algebra, from the 16th century, when cubic and quartic equations were solved, until the first half of the 19th century, when the impossibility of such a general solution was proved (Abel–Ruffini theorem).

ALGEBRAIC EQUATIONS

An algebraic equation in n variables is an polynomial equation of the form.

$$f\left(x_1, x_2, \ldots, x_n\right) = \sum_{e_1, \ldots, e_n} c_{e_1,\ e_2, \ldots, e_n} x_1^{e_1} x_2^{e_2} \ldots x_n^{e_n} = 0,$$

where the coefficients $c_{e_1,\ e_2, \ldots, e_n}$ are integers (where the exponents e_i are nonnegative integers and the sum is finite).

Examples of algebraic equations are given in the following table.

Curve	Equation
Cayley's sextic	$4\left(x^2 + y^2 - x\right)^3 - 27\left(x^2 + y^2\right)^2 = 0$
Eight curve	$x^4 - \left(x^2 - y^2\right) = 0$
Line through $(1,0)$ and $(0,1)$	$x + y - 1 = 0$

Plane through $(1,0,0)$, $(0,1,0)$, and $(0,0,1)$	$x+y+z-1=0$
Unit circle	$x^2-y^2-1=0$
Unit sphere	$x^2-y^2+z^2-1=0$

The roots of an algebraic equation in one variable are known as algebraic numbers.

Solving Algebraic Equations

For theoretical work and applications one often needs to find numbers that, when substituted for the unknown, make a certain polynomial equal to zero. Such a number is called a "root" of the polynomial. For example, the polynomial.

$-16t^2 + 88t + 48$

represents the height above Earth at t seconds of a projectile thrown straight up at 88 feet per second from the top of a tower 48 feet high. (The 16 in the formula comes from one-half the acceleration of gravity, 32 feet per second per second.) By setting the equation equal to zero and factoring it as $(4t - 24)(-4t - 2) = 0$, the equation's one positive root is found to be 6, meaning that the object will hit the ground about 6 seconds after it is thrown. (This problem also illustrates the important algebraic concept of the zero factor property: if $ab = 0$, then either $a = 0$ or $b = 0$).

The theorem that every polynomial has as many complex roots as its degree is known as the fundamental theorem of algebra and was first proved in 1799 by the German mathematician Carl Friedrich Gauss. Simple formulas exist for finding the roots of the general polynomials of degrees one and two, and much less simple formulas exist for polynomials of degrees three and four. The French mathematician Évariste Galois discovered, shortly before his death in 1832, that no such formula exists for a general polynomial of degree greater than four. Many ways exist, however, of approximating the roots of these polynomials.

Liner and quadratic formulas
$ax+b=0 \Rightarrow x=\dfrac{-b}{a}$
$ax^2+bx+c=0 \Rightarrow x=\dfrac{-b\pm\sqrt{b^2-4ac}}{2a}$

Solving Systems of Algebraic Equations

An extension of the study of single equations involves multiple equations that are solved simultaneously—so-called systems of equations. For example, the intersection of two straight lines, $ax + by = c$ and $Ax + By = C$, can be found algebraically by discovering the values of x and y that simultaneously solve each equation. The earliest systematic development of methods for solving systems of equations occurred in ancient China. An adaptation of a problem from the 1st-century-AD Chinese classic *Nine Chapters on the Mathematical Procedures* illustrates how such systems arise.

Imagine there are two kinds of wheat and that you have four sheaves of the first type and five sheaves of the second type. Although neither of these is enough to produce a bushel of wheat, you can produce a bushel by adding three sheaves of the first type to five of the second type, or you can produce a bushel by adding four sheaves of the first type to two of the second type. What fraction of a bushel of wheat does a sheaf of each type of wheat contain?

Using modern notation, suppose we have two types of wheat, respectively, and x and y represent the number of bushels obtained per sheaf of the first and second types, respectively. Then the problem leads to the system of equations:

$3x + 5y = 1$ (bushel)

$4x + 2y = 1$ (bushel)

A simple method for solving such a system is first to solve either equation for one of the variables. For example, solving the second equation for y yields $y = 1/2 - 2x$. The right side of this equation can then be substituted for y in the first equation ($3x + 5y = 1$), and then the first equation can be solved to obtain x (= 3/14). Finally, this value of x can be substituted into one of the earlier equations to obtain y (= 1/14). Thus, the first type yields 3/14 bushels per sheaf and the second type yields 1/14. Note that the solution (3/14, 1/14) would be difficult to discern by graphing techniques. In fact, any precise value based on a graphing solution may be only approximate; for example, the point (0.0000001, 0) might look like (0, 0) on a graph, but even such a small difference could have drastic consequences in the real world.

Rather than individually solving each possible system of two equations in two unknowns, the general system can be solved. To return to the general equations given above:

$ax + by = c$

$Ax + By = C$

The solutions are given by $x = (Bc - bC)/(aB - Ab)$ and $y = (Ca - cA)/(aB - Ab)$. Note that the denominator of each solution, $(aB - Ab)$, is the same. It is called the determinant of the system, and systems in which the denominator is equal to zero have either no solution (in which case the equations represent parallel lines) or infinitely many solutions (in which case the equations represent the same line).

One can generalize simultaneous systems to consider m equations in n unknowns. In this case, one usually uses subscripted letters x_1, x_2, ..., xn for the unknowns and $a_{1,1}$, ..., $a_{1,n}$; $a_{2,1}$, ..., $a_{2,n}$; ...; $a_{m,1}$, ..., $a_{m,n}$ for the coefficients of each equation, respectively. When $n = 3$ one is dealing with planes in three-dimensional space, and for higher values of n one is dealing with hyperplanes in spaces of higher dimension. In general, n equations in m unknowns have infinitely many solutions when $m < n$ and no solutions when $m > n$. The case $m = n$ is the only case where there can exist a unique solution.

LINEAR EQUATION

In mathematics, a linear equation is an equation that may be put in the form:

$$a_1 x_1 + \cdots + a_n x_n + b = 0,$$

where $x_1, \ldots, x_n$ are the variables (or unknowns or indeterminates), and $b, a_1, \ldots, a_n$ are the coefficients, which are often real numbers. The coefficients may be considered as parameters of the equation, and may be arbitrary expressions, provided they do not contain any of the variables. To yield a meaningful equation, the coefficient s $a_1, \ldots, a_n$ are required to not be all zero.

In other words, a linear equation is obtained by equating to zero a linear polynomial over some field, from which the coefficients are taken (the symbols used for the variables are supposed to not denote any element of the field).

The solutions of such an equation are the values that, when substituted for the unknowns, make the equality true.

In the case of just one variable, there is exactly one solution (provided that $a_1 \neq 0$). Often, the term *linear equation* refers implicitly to this particular case, in which the variable is sensibly called the *unknown*.

In the case of two variables, each solution may be interpreted as the Cartesian coordinates of a point of the Euclidean plane. The solutions of a linear equation form a line in the Euclidean plane, and, conversely, every line can be viewed as the set of all solutions of a linear equation in two variables. This is the origin of the term *linear* for describing this type of equations. More generally, the solutions of a linear equation in n variables form a hyperplane (a subspace of dimension $n - 1$) in the Euclidean space of dimension n.

Linear equations occur frequently in all mathematics and their applications in physics and engineering, partly because non-linear systems are often well approximated by linear equations.

One Variable

Frequently the term *linear equation* refers implicitly to the case of just one variable.

In this case, the equation can be put in the form

$$ax + b = 0,$$

and it has a unique solution

$$x = -\frac{b}{a}$$

in the general case where $a \neq 0$. In this case, the name *unknown* is sensibly given to the variable x.

If $a = 0$, there are two cases. Either b equals also 0, and every number is a solution. Otherwise $b \neq 0$, and there is no solution. In this latter case, the equation is said to be inconsistent.

Two Variables

In the case of two variables, any linear equation can be put in the form

$$ax + by + c = 0,$$

where the variables are x and y, and the coefficients are a, b and c.

An equivalent equation (that is an equation with exactly the same solutions) is

$$Ax + By = C,$$

with $A = a$, $B = b$, and $C = -c$.

These equivalent variants are sometimes given generic names, such as *general form* or *standard form*.

There are other forms for a linear equation, which can all be transformed in the standard form with simple algebraic manipulations, such as adding the same quantity to both members of the equation, or multiplying both members by the same nonzero constant.

Linear Function

If $b \neq 0$, the equation

$$ax + by + c = 0$$

is a linear equation in the single variable y for every value of x. It has therefore a unique solution for y, which is given by

$$y = -\frac{a}{b}x - \frac{c}{b}.$$

This defines a function. The graph of this function is a line with slope $-\frac{a}{b}$ and y-intercept $-\frac{c}{b}$.

The functions whose graph is a line are generally called *linear functions* in the context of calculus. However, in linear algebra, a linear function is a function that maps a sum to the sum of the images of the summands. So, for this definition, the above function is linear only when $c = 0$, that is when the line passes through the origin. For avoiding confusion, the functions whose graph is an arbitrary line are often called *affine functions*.

Geometric Interpretation

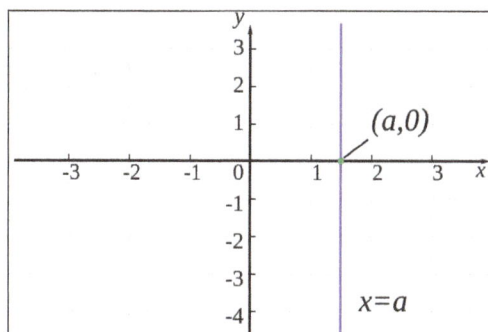

Vertical line of equation $x = a$.

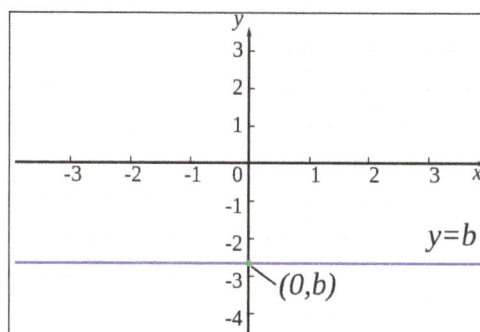

Horizontal line of equation $y = b$.

Each solution (x, y) of a linear equation

$$ax + by + c = 0$$

may be viewed as the Cartesian coordinates of a point in the Euclidean plane. With this interpretation, all solutions of the equation form a line, provided that a and b are not both zero. Conversely, every line is the set of all solutions of a linear equation.

The phrase "linear equation" takes its origin in this correspondence between lines and equations: a *linear equation* in two variables is an equation whose solutions form a line.

If $b \neq 0$, the line is the graph of the function of x that has been defined in the preceding section. If $b = 0$, the line is a *vertical line* (that is a line parallel to the y-axis) of equation $x = -\dfrac{c}{a}$, which is not the graph of a function of x.

Similarly, if $a \neq 0$, the line is the graph of a function of y, and, if $a = 0$, one has a horizontal line of equation $y = -\dfrac{c}{b}$.

Equation of a Line

There are various ways of defining a line. In the following subsections, a linear equation of the line is given in each case.

Slope–intercept Form

A non-vertical line can be defined by its slope (mathematics) m, and its y-intercept y_0 (the y coordinate of its intersection with the y-axis). In this case its *linear equation* can be written

$$y = mx + y_0.$$

If, moreover, the line is not horizontal, it can be defined by its slope and its x-intercept x_0. In this case, its equation can be written

$$y = m(x - x_0),$$

or, equivalently,

$$y = mx - mx_0.$$

These form rely on the habit of considering a non vertical line as the graph of a function. For a line given by an equation

$$ax + by + c = 0$$

these forms can be easily deduced from the relations

$$m = -\frac{a}{b},$$

$$x_0 = -\frac{c}{a},$$

$$y_0 = -\frac{c}{b}.$$

Point–slope Form

A non-vertical line can be defined by its slope (mathematics) m, and the coordinates x_1, y_1 of any point of the line. In this case, a linear equation of the line is

$$y = y_1 + m(x - x_1),$$

or

$$y = mx + y_1 - mx_1.$$

This equation can also be written

$$y - y_1 = m(x - x_1)$$

for emphasizing that the slope of a line can be computed from the coordinates of any two points.

Intercept Form

A line that is not parallel to an axis and does not passes through origin cuts the axes in two different points. The intercept values x_0 and y_0 of these two points are nonzero, and an equation of the line is

$$\frac{x}{x_0} + \frac{y}{y_0} = 1.$$

(It easy to verify that the line defined by this equation has x_0 and y_0 as intercept values).

Two-point Form

Given two different points (x_1, y_1) and (x_2, y_2), there is exactly one line that passes through them. There are several ways to write a linear equation of is line.

If $x_1 \neq x_2$, the slope of the line is $\dfrac{y_2 - y_1}{x_2 - x_1}$. Thus, a point-slope form is

$$y - y_1 = \frac{y_2 - y_1}{x_2 - x_1}(x - x_1).$$

By clearing denominators, one gets the equation

$$(x_2 - x_1)(y - y_1) - (y_2 - y_1)(x - x_1) = 0,$$

which is valid also when $x_1 = x_2$ (for verifying this, it suffices to verify that the two given points satisfy the equation).

This form is not symmetric in the two given points, but a symmetric form can be obtained by regrouping the constant terms:

$$(y_1 - y_2)x + (x_2 - x_1)y + (x_1 y_2 - x_2 y_1) = 0$$

(exchanging the two points changes the sign of the left-hand side of the equation).

Determinant Form

The two-point form of the equation of a line can be expressed simply in terms of a determinant. There are two common ways for that.

The equation $(x_2 - x_1)(y - y_1) - (y_2 - y_1)(x - x_j) = 0$ is the result of expanding the determinant in the equation

$$\begin{vmatrix} x - x_1 & y - y_1 \\ x_2 - x_1 & y_2 - y_1 \end{vmatrix} = 0.$$

The equation $(y_1 - y_2)x + (x_2 - x_1)y + (x_1 y_2 - x_2 y_1) = 0$ can be obtained be expanding with respect to its first row the determinant in the equation

$$\begin{vmatrix} x & y & 1 \\ x_1 & y_1 & 1 \\ x_2 & y_2 & 1 \end{vmatrix} = 0.$$

Beside being very simple and mnemonic, this form has the advantage of being a special case of the more general equation of a hyperplane passing through n points in a space of dimension $n - 1$. These equations rely on the condition of linear dependence of points in a projective space.

More than two Variables

A linear equation with more than two variables may always be assumed to have the form

$$a_1 x_1 + a_2 x_2 + \cdots + a_n x_n + b = 0.$$

The coefficient b, often denoted a_0 is called the *constant term*, sometimes the *absolute term*. Depending of the context, the term *coefficient* can be reserved for the a_i with $i > 0$.

When dealing with $n = 3$ variables, it is common to use x, y and z instead of indexed variables.

A solution of such an equation is a n-tuples such that substituting each element of the tuple for the corresponding variable transforms the equation into a true equality.

For an equation to be meaningful, the coefficient of at least one variable must be non-zero. In fact, if every variable has a zero coefficient, then, as mentioned for one variable, the equation is either *inconsistent* (for $b \neq 0$) as having no solution, or all n-tuples are solutions.

The n-tuples that are solutions of a linear equation in n variables are the Cartesian coordinates of the points of an $(n - 1)$-dimensional hyperplane in an n-dimensional Euclidean space (or affine space if the coefficients are complex numbers or belong to any field). In the case of three variable, this hyperplane is a plane.

If a linear equation is given with $a_j \neq 0$, then the equation can be solved for x_j, yelding

$$x_j = -\frac{b}{a_j} - \sum_{i \in \{1,\ldots,n\}, i \neq j} \frac{a_i}{a_j} x_i.$$

If the coefficients are real numbers, this defines a real-valued function of n real variables.

QUADRATIC EQUATION

$$x = \frac{-b \pm \sqrt{b^2 - 4ac}}{2a}$$

The quadratic formula for the roots of the general quadratic equation.

In algebra, a quadratic equation is any equation having the form

$$ax^2 + bx + c = 0,$$

where x represents an unknown, and a, b, and c represent known numbers, with $a \neq 0$. If $a = 0$, then the equation is linear, not quadratic, as there is no ax^2 term. The numbers a, b, and c are the *coefficients* of the equation and may be distinguished by calling them, respectively, the *quadratic coefficient*, the *linear coefficient* and the *constant* or *free term*.

The values of x that satisfy the equation are called *solutions* of the equation, and *roots* or *zeros* of its left-hand side. A quadratic equation has at most two solutions. If there is no real solution, there are two complex solutions. If there is only one solution, one says that it is a double root. So a quadratic equation has always two roots, if complex roots are considered, and if a double root is counted for two. If the two solutions are denoted r and s (possibly equal), one has

$$ax^2 + bx + c = a(x - r)(x - s).$$

Thus, the process of solving a quadratic equation is also called *factorizing* or *factoring*. Completing the square is the standard method for that, which results in the quadratic formula, which express the solutions in terms of a, b, and c. Graphing may also be used for getting an approximate value of the solutions. Solutions to problems that may be expressed in terms of quadratic equations were known as early as 2000 BC.

Because the quadratic equation involves only one unknown, it is called "univariate". The quadratic equation only contains powers of x that are non-negative integers, and therefore it is a polynomial equation. In particular, it is a second-degree polynomial equation, since the greatest power is two.

Solving the Quadratic Equation

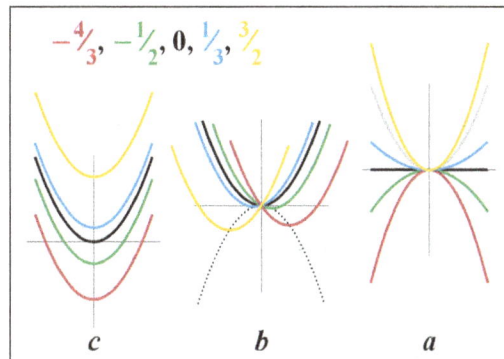

Plots of quadratic function $y = ax^2 + bx + c$, varying each coefficient separately while the other coefficients are fixed (at values $a = 1$, $b = 0$, $c = 0$).

A quadratic equation with real or complex coefficients has two solutions, called *roots*. These two solutions may or may not be distinct, and they may or may not be real.

Factoring by Inspection

It may be possible to express a quadratic equation $ax^2 + bx + c = 0$ as a product $(px + q)(rx + s) = 0$. In some cases, it is possible, by simple inspection, to determine values of p, q, r, and s that make the two forms equivalent to one another. If the quadratic equation is written in the second form, then the "Zero Factor Property" states that the quadratic equation is satisfied if $px + q = 0$ or $rx + s = 0$. Solving these two linear equations provides the roots of the quadratic.

For most students, factoring by inspection is the first method of solving quadratic equations to which they are exposed. If one is given a quadratic equation in the form $x^2 + bx + c = 0$, the sought factorization has the form $(x + q)(x + s)$, and one has to find two numbers q and s that add up to b and whose product is c (this is sometimes called "Vieta's rule" and is related to Vieta's formulas). As an example, $x^2 + 5x + 6$ factors as $(x + 3)(x + 2)$. The more general case where a does not equal 1 can require a considerable effort in trial and error guess-and-check, assuming that it can be factored at all by inspection.

Except for special cases such as where $b = 0$ or $c = 0$, factoring by inspection only works for quadratic equations that have rational roots. This means that the great majority of quadratic equations that arise in practical applications cannot be solved by factoring by inspection.

Completing the Square

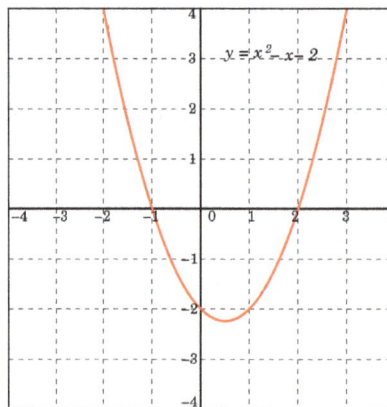

For the quadratic function $y = x^2 - x - 2$, the points where the graph crosses the x-axis, $x = -1$ and $x = 2$, are the solutions of the quadratic equation $x^2 - x - 2 = 0$.

The process of completing the square makes use of the algebraic identity,

$$x^2 + 2hx + h^2 = (x+h)^2,$$

which represents a well-defined algorithm that can be used to solve any quadratic equation. Starting with a quadratic equation in standard form, $ax^2 + bx + c = 0$.

1. Divide each side by a, the coefficient of the squared term.

2. Subtract the constant term c/a from both sides.

3. Add the square of one-half of b/a, the coefficient of x, to both sides. This "completes the square", converting the left side into a perfect square.

4. Write the left side as a square and simplify the right side if necessary.

5. Produce two linear equations by equating the square root of the left side with the positive and negative square roots of the right side.

6. Solve the two linear equations.

We illustrate use of this algorithm by solving $2x^2 + 4x - 4 = 0$

$$1)\ x^2 + 2x - 2 = 0$$

$$2)\ x^2 + 2x = 2$$

$$3)\ x^2 + 2x + 1 = 2 + 1$$

$$4)\ (x+1)^2 = 3$$

$$5)\ x + 1 = \pm\sqrt{3}$$

$$6)\ x = -1 \pm \sqrt{3}$$

The plus-minus symbol "$\pm$" indicates that both $x = -1 + \sqrt{3}$ and $x = -1 - \sqrt{3}$ are solutions of the quadratic equation.

Quadratic Formula and its Derivation

Completing the square can be used to derive a general formula for solving quadratic equations, called the quadratic formula. The mathematical proof will now be briefly summarized. It can easily be seen, by polynomial expansion, that the following equation is equivalent to the quadratic equation:

$$\left(x + \frac{b}{2a}\right)^2 = \frac{b^2 - 4ac}{4a^2}.$$

Taking the square root of both sides, and isolating x, gives:

$$x = \frac{-b \pm \sqrt{b^2 - 4ac}}{2a}.$$

Some sources, particularly older ones, use alternative parameterizations of the quadratic equation such as $ax^2 + 2bx + c = 0$ or $ax^2 - 2bx + c = 0$, where b has a magnitude one half of the more common one, possibly with opposite sign. These result in slightly different forms for the solution, but are otherwise equivalent.

A number of alternative derivations can be found in the literature. These proofs are simpler than the standard completing the square method, represent interesting applications of other frequently used techniques in algebra, or offer insight into other areas of mathematics.

A lesser known quadratic formula, as used in Muller's method provides the same roots via the equation:

$$x = \frac{2c}{-b \pm \sqrt{b^2 - 4ac}}.$$

This can be deduced from the standard quadratic formula by Vieta's formulas, which assert that the product of the roots is c/a.

One property of this form is that it yields one valid root when $a = 0$, while the other root contains division by zero, because when $a = 0$, the quadratic equation becomes a linear equation, which has one root. By contrast, in this case, the more common formula has a division by zero for one root and an indeterminate form $0/0$ for the other root. On the other hand, when $c = 0$, the more common formula yields two correct roots whereas this form yields the zero root and an indeterminate form $0/0$.

Reduced Quadratic Equation

It is sometimes convenient to reduce a quadratic equation so that its leading coefficient is one. This is done by dividing both sides by a, which is always possible since a is non-zero. This produces the *reduced quadratic equation*:

$$x^2 + px + q = 0,$$

where $p = b/a$ and $q = c/a$. This monic equation has the same solutions as the original.

The quadratic formula for the solutions of the reduced quadratic equation, written in terms of its coefficients, is:

$$x = \frac{1}{2}\left(-p \pm \sqrt{p^2 - 4q}\right),$$

or equivalently:

$$x = -\frac{p}{2} \pm \sqrt{\left(\frac{p}{2}\right)^2 - q}.$$

Discriminant

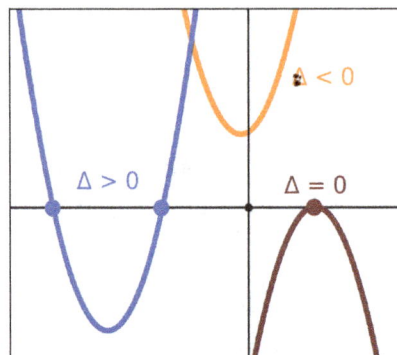

Discriminant signs.

In the quadratic formula, the expression underneath the square root sign is called the *discriminant* of the quadratic equation, and is often represented using an upper case D or an upper case Greek delta:

$$\Delta = b^2 - 4ac.$$

A quadratic equation with *real* coefficients can have either one or two distinct real roots, or two distinct complex roots. In this case the discriminant determines the number and nature of the roots. There are three cases:

- If the discriminant is positive, then there are two distinct roots

  $$\frac{-b+\sqrt{\Delta}}{2a} \quad \text{and} \quad \frac{-b-\sqrt{\Delta}}{2a},$$

  both of which are real numbers. For quadratic equations with rational coefficients, if the discriminant is a square number, then the roots are rational—in other cases they may be quadratic irrationals.

- If the discriminant is zero, then there is exactly one real root

  $$-\frac{b}{2a},$$

  sometimes called a repeated or double root.

- If the discriminant is negative, then there are no real roots. Rather, there are two distinct (non-real) complex roots

  $$\frac{b}{2a} + i\frac{\sqrt{-\ddot{A}}}{2a} \quad \text{and} \quad -\frac{b}{2a} - i\frac{\sqrt{-\ddot{A}}}{2a},$$

  which are complex conjugates of each other. In these expressions i is the imaginary unit.

Thus the roots are distinct if and only if the discriminant is non-zero, and the roots are real if and only if the discriminant is non-negative.

Geometric Interpretation

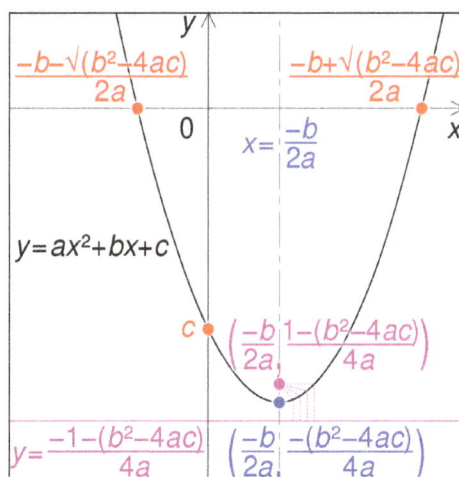

Graph of $y = ax^2 + bx + c$, where a and the discriminant $b^2 - 4ac$ are positive, with

- Roots and y-intercept in red.

- Vertex and axis of symmetry in blue.

- Focus and directrix in pink.

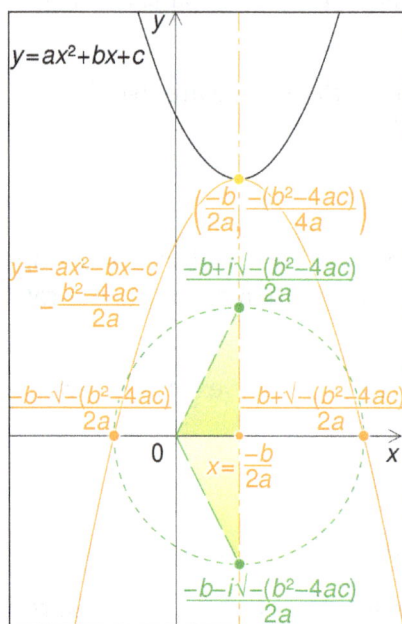

Visualisation of the complex roots of $y = ax^2 + bx + c$: the parabola is rotated 180° about its vertex (orange). Its x-intercepts are rotated 90° around their mid-point, and the Cartesian plane is interpreted as the complex plane (green).

The function $f(x) = ax^2 + bx + c$ is the quadratic function. The graph of any quadratic function has the same general shape, which is called a parabola. The location and size of the parabola, and how it opens, depend on the values of a, b, and c. As shown in figure, if $a > 0$, the parabola has a minimum point and opens upward. If $a < 0$, the parabola has a maximum point and opens downward. The extreme point of the parabola, whether minimum or maximum, corresponds to its vertex. The *x-coordinate* of the vertex will be located at $x = \frac{-b}{2a}$, and the *y-coordinate* of the vertex may be found by substituting this *x-value* into the function. The *y-intercept* is located at the point (0, c).

The solutions of the quadratic equation $ax^2 + bx + c = 0$ correspond to the roots of the function $f(x) = ax^2 + bx + c$, since they are the values of x for which $f(x) = 0$. As shown in figure, if a, b, and c are real numbers and the domain of f is the set of real numbers, then the roots of f are exactly the x-coordinates of the points where the graph touches the x-axis. As shown in figure, if the discriminant is positive, the graph touches the x-axis at two points; if zero, the graph touches at one point; and if negative, the graph does not touch the x-axis.

Quadratic Factorization

The term

$$x - r$$

is a factor of the polynomial

$$ax^2 + bx + c$$

if and only if r is a root of the quadratic equation

$$ax^2 + bx + c = 0.$$

It follows from the quadratic formula that

$$ax^2 + bx + c = a\left(x - \frac{-b + \sqrt{b^2 - 4ac}}{2a}\right)\left(x - \frac{-b - \sqrt{b^2 - 4ac}}{2a}\right).$$

In the special case $b^2 = 4ac$ where the quadratic has only one distinct root (*i.e.* the discriminant is zero), the quadratic polynomial can be factored as

$$ax^2 + bx + c = a\left(x + \frac{b}{2a}\right)^2.$$

Graphical Solution

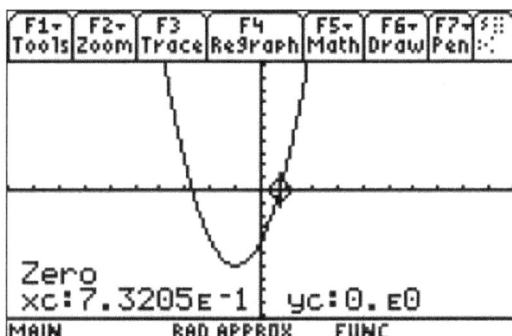

Graphing calculator computation of one of the two roots of the quadratic equation $2x^2 + 4x - 4 = 0$. Although the display shows only five significant figures of accuracy, the retrieved value of xc is 0.732050807569, accurate to twelve significant figures.

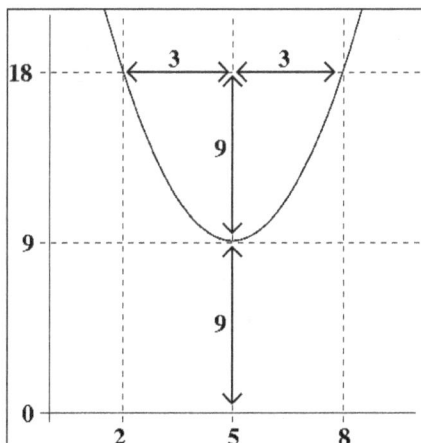

A quadratic function without real root: $y = (x - 5)^2 + 9$. The "3" is the imaginary part of the x-intercept. The real part is the x-coordinate of the vertex. Thus the roots are $5 \pm 3i$.

The solutions of the quadratic equation

$$ax + bx + c =$$

may be deduced from the graph of the quadratic function

$$y = ax^2 + bx + c,$$

which is a parabola.

If the parabola intersects the x-axis in two points, there are two real roots, which are the x-coordinates of these two points (also called x-intercept).

If the parabola is tangent to the x-axis, there is a double root, which is the x-coordinate of the contact point between the graph and parabola.

If the parabola does not intersect the x-axis, there are two complex conjugate roots. Although these roots cannot be visualized on the graph, their real and imaginary parts can be.

Let h and k be respectively the x-coordinate and the y-coordinate of the vertex of the parabola (that is the point with maximal or minimal y-coordinate. The quadratic function may be rewritten

$$y = a(x-h)^2 + k.$$

Let d be the distance between the point of y-coordinate $2k$ on the axis of the parabola, and a point on the parabola with the same y-coordinate. Then the real part of the roots is h, and their imaginary part are $\pm d$. That is, the roots are

$$h + id \quad \text{and} \quad x - id,$$

or in the case of the example of the figure

$$5 + 3i \quad \text{and} \quad 5 - 3i.$$

Avoiding Loss of Significance

Although the quadratic formula provides an exact solution, the result is not exact if real numbers are approximated during the computation, as usual in numerical analysis, where real numbers are approximated by floating point numbers (called "reals" in many programming languages). In this context, the quadratic formula is not completely stable.

This occurs when the roots have different order of magnitude, or, equivalently, when b^2 and $b^2 - 4ac$ are close in magnitude. In this case, the subtraction of two nearly equal numbers will cause loss of significance or catastrophic cancellation in the smaller root. To avoid this, the root that is smaller in magnitude, r, can be computed as $(c/a)/R$ where R is the root that is bigger in magnitude.

A second form of cancellation can occur between the terms b^2 and $4ac$ of the discriminant, that is when the two roots are very close. This can lead to loss of up to half of correct significant figures in the roots.

Examples and Applications

The trajectory of the cliff jumper is parabolic because horizontal displacement is a linear function of time $x = v_x t$, while vertical displacement is a quadratic function of time $y = \frac{1}{2}at^2 + v_y t + h$. As a result, the path follows quadratic equation $y = \frac{a}{2v_x^2}x^2 + \frac{v_y}{v_x}x + h$, where v_x and v_y are horizontal and vertical components of the original velocity, a is gravitational acceleration and h is original height. The a value should be considered negative here, as its direction (downwards) is opposite to the height measurement (upwards).

The golden ratio is found as the positive solution of the quadratic equation $x^2 - x - 1 = 0$.

The equations of the circle and the other conic sections—ellipses, parabolas, and hyperbolas—are quadratic equations in two variables.

Given the cosine or sine of an angle, finding the cosine or sine of the angle that is half as large involves solving a quadratic equation.

The process of simplifying expressions involving the square root of an expression involving the square root of another expression involves finding the two solutions of a quadratic equation.

Descartes' theorem states that for every four kissing (mutually tangent) circles, their radii satisfy a particular quadratic equation.

The equation given by Fuss' theorem, giving the relation among the radius of a bicentric quadrilateral's inscribed circle, the radius of its circumscribed circle, and the distance between the centers of those circles, can be expressed as a quadratic equation for which the distance between the two circles' centers in terms of their radii is one of the solutions. The other solution of the same equation in terms of the relevant radii gives the distance between the circumscribed circle's center and the center of the excircle of an ex-tangential quadrilateral.

Alternative Methods of Root Calculation

Vieta's Formulas

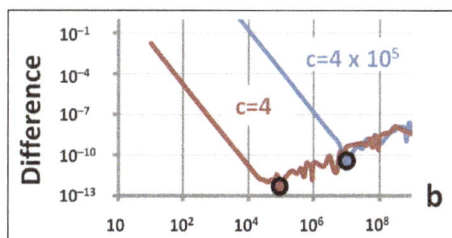

Graph of the difference between Vieta's approximation for the smaller of the two roots of the quadratic equation $x^2 + bx + c = 0$ compared with the value calculated using the quadratic formula. Vieta's approximation is inaccurate for small b but is accurate for large b. The direct evaluation using the quadratic formula is accurate for small b with roots of comparable value but experiences loss of significance errors for large b and widely spaced roots. The difference between Vieta's approximation *versus* the direct computation reaches a minimum at the large dots, and rounding causes squiggles in the curves beyond this minimum.

Vieta's formulas give a simple relation between the roots of a polynomial and its coefficients. In the case of the quadratic polynomial, they take the following form:

$$x_1 + x_2 = -\frac{b}{a}$$

and

$$x_1 x_2 = \frac{c}{a}.$$

These results follow immediately from the relation:

$$(x - x_1)(x - x_2) = x^2 - (x_1 + x_2)x + x_1 x_2 = 0,$$

which can be compared term by term with

$$x^2 + (b/a)x + c/a = 0.$$

The first formula above yields a convenient expression when graphing a quadratic function. Since the graph is symmetric with respect to a vertical line through the vertex, when there are two real roots the vertex's x-coordinate is located at the average of the roots (or intercepts). Thus the x-coordinate of the vertex is given by the expression,

$$x_V = \frac{x_1 + x_2}{2} = -\frac{b}{2a}.$$

The y-coordinate can be obtained by substituting the above result into the given quadratic equation, giving,

$$y_V = -\frac{b^2}{4a} + c = -\frac{b^2 - 4ac}{4a}.$$

As a practical matter, Vieta's formulas provide a useful method for finding the roots of a quadratic in the case where one root is much smaller than the other. If $|x_2| \ll |x_1|$, then $x_1 + x_2 \approx x_1$, and we have the estimate:

$$x_1 \approx -\frac{b}{a}.$$

The second Vieta's formula then provides:

$$x_2 = \frac{c}{a x_1} \approx -\frac{c}{b}.$$

These formulas are much easier to evaluate than the quadratic formula under the condition of one large and one small root, because the quadratic formula evaluates the small root as the difference of two very nearly equal numbers (the case of large b), which causes round-off error in a numerical evaluation. As the linear coefficient b increases, initially the quadratic formula is accurate, and the approximate formula improves in accuracy, leading to a smaller difference between the methods as b increases. However, at some point the quadratic formula begins to lose accuracy because of round off error, while the approximate method continues to improve. Consequently, the difference between the methods begins to increase as the quadratic formula becomes worse and worse.

This situation arises commonly in amplifier design, where widely separated roots are desired to ensure a stable operation.

Trigonometric Solution

In the days before calculators, people would use mathematical tables—lists of numbers showing the results of calculation with varying arguments—to simplify and speed up computation. Specialized tables were published for applications such as astronomy, celestial navigation and statistics. Methods of numerical approximation existed, called prosthaphaeresis, that offered shortcuts around time-consuming operations such as multiplication and taking powers and roots. Astronomers, especially, were concerned with methods that could speed up the long series of computations involved in celestial mechanics calculations.

It is within this context that we may understand the development of means of solving quadratic equations by the aid of trigonometric substitution. Consider the following alternate form of the quadratic equation,

$$ax^2 + bx \pm c = 0,$$

where the sign of the $\pm$ symbol is chosen so that a and c may both be positive. By substituting

$$x \quad \sqrt{c/a}\, \tan$$

and then multiplying through by $\cos^2 \theta$, we obtain

$$\sin^2 \theta + \frac{b}{\sqrt{ac}} \sin \theta \cos \theta \pm \cos^2 \theta = 0.$$

Introducing functions of 2θ and rearranging, we obtain

$$\tan 2\theta_n = +2 \frac{\sqrt{ac}}{b},$$

$$\sin 2\theta_p = -2 \frac{\sqrt{ac}}{b},$$

where the subscripts n and p correspond, respectively, to the use of a negative or positive sign in equation $ax^2 + bx \pm c = 0$. Substituting the two values of θ_n or θ_p found from equations $\tan 2\theta_n = +2 \frac{\sqrt{ac}}{b}$, or $\sin 2\theta_p = -2 \frac{\sqrt{ac}}{b}$, into $x = \sqrt{c/a}\, \tan \theta$ gives the required roots of

$ax^2 + bx \pm c = 0$. Complex roots occur in the solution based on equation if the absolute value of $\sin 2\theta_p$ exceeds unity. The amount of effort involved in solving quadratic equations using this mixed trigonometric and logarithmic table look-up strategy was two-thirds the effort using logarithmic tables alone. Calculating complex roots would require using a different trigonometric form.

To illustrate, let us assume we had available seven-place logarithm and trigonometric tables, and wished to solve the following to six-significant-figure accuracy:

$$4.16130x^2 + 9.15933x - 11.4207 = 0$$

1. A seven-place lookup table might have only 100,000 entries, and computing intermediate results to seven places would generally require interpolation between adjacent entries.

2. $\log a = 0.6192290, \log b = 0.9618637, \log c = 1.0576927$

3. $2\sqrt{ac}/b = 2 \times 10^{(0.6192290+1.0576927)/2 - 0.9618637} = 1.505314$

4. $\theta = (\tan^{-1} 1.505314)/2 = 28.20169° \text{ or } -61.79831°$

5. $\log|\tan\theta| = -0.2706462 \text{ or } 0.2706462$

6. $\log\sqrt{c/a} = (1.0576927 - 0.6192290)/2 = 0.2192318$

7. $x_1 = 10^{0.2192318 - 0.2706462} = 0.888353$ (rounded to six significant figures)

   $x_2 = -10^{0.2192318 + 0.2706462} = -3.08943$

Solution for Complex Roots in Polar Coordinates

If the quadratic equation $ax^2 + bx + c = 0$ with real coefficients has two complex roots—the case where $b^2 - 4ac < 0$, requiring a and c to have the same sign as each other—then the solutions for the roots can be expressed in polar form as

$$x_1, x_2 = r(\cos\theta \pm i\sin\theta),$$

where $r = \sqrt{\frac{c}{a}}$ and $\theta = \cos^{-1}\left(\frac{-b}{2\sqrt{ac}}\right)$.

Geometric Solution

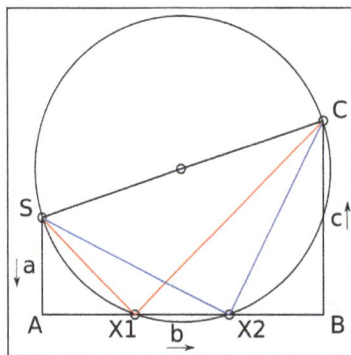

Geometric solution of $ax^2 + bx + c = 0$ using Lill's method.
Solutions are −AX1/SA, −AX2/SA.

The quadratic equation may be solved geometrically in a number of ways. One way is via Lill's method. The three coefficients a, b, c are drawn with right angles between them as in SA, AB, and BC in Figure . A circle is drawn with the start and end point SC as a diameter. If this cuts the middle line AB of the three then the equation has a solution, and the solutions are given by negative of the distance along this line from A divided by the first coefficient a or SA. If a is 1 the coefficients may be read off directly. Thus the solutions in the diagram are −AX1/SA and −AX2/SA.

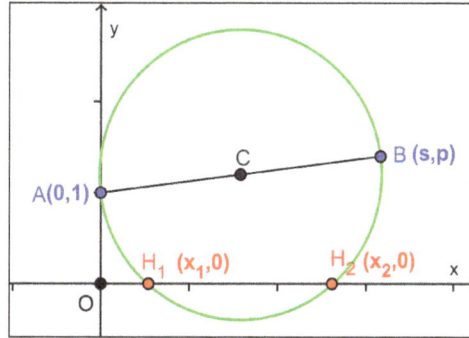

Carlyle circle of the quadratic equation $x^2 - sx + p = 0$.

The Carlyle circle, named after Thomas Carlyle, has the property that the solutions of the quadratic equation are the horizontal coordinates of the intersections of the circle with the horizontal axis. Carlyle circles have been used to develop ruler-and-compass constructions of regular polygons.

Generalization of Quadratic Equation

The formula and its derivation remain correct if the coefficients a, b and c are complex numbers, or more generally members of any field whose characteristic is not 2. (In a field of characteristic 2, the element $2a$ is zero and it is impossible to divide by it).

The symbol,

$$\pm\sqrt{b^2 - 4ac}$$

in the formula should be understood as "either of the two elements whose square is $b^2 - 4ac$, if such elements exist". In some fields, some elements have no square roots and some have two; only zero has just one square root, except in fields of characteristic 2. Even if a field does not contain a square root of some number, there is always a quadratic extension field which does, so the quadratic formula will always make sense as a formula in that extension field.

Characteristic 2

In a field of characteristic 2, the quadratic formula, which relies on 2 being a unit, does not hold. Consider the monic quadratic polynomial,

$$x^2 + bx + c$$

over a field of characteristic 2. If $b = 0$, then the solution reduces to extracting a square root, so the solution is,

$$x = \sqrt{c}$$

and there is only one root since,

$$-\sqrt{c} = -\sqrt{c} + 2\sqrt{c} = \sqrt{c}.$$

In summary,

$$x^2 + c = (x + \sqrt{c})^2.$$

In the case that $b \neq 0$, there are two distinct roots, but if the polynomial is irreducible, they cannot be expressed in terms of square roots of numbers in the coefficient field. Instead, define the 2-root $R(c)$ of c to be a root of the polynomial $x^2 + x + c$, an element of the splitting field of that polynomial. One verifies that $R(c) + 1$ is also a root. In terms of the 2-root operation, the two roots of the (non-monic) quadratic $ax^2 + bx + c$ are,

$$\frac{b}{a} R\left(\frac{ac}{b^2}\right)$$

and

$$\frac{b}{a}\left(R\left(\frac{ac}{b^2}\right) + 1\right).$$

For example, let a denote a multiplicative generator of the group of units of F_4, the Galois field of order four (thus a and $a + 1$ are roots of $x^2 + x + 1$ over F_4. Because $(a + 1)^2 = a$, $a + 1$ is the unique solution of the quadratic equation $x^2 + a = 0$. On the other hand, the polynomial $x^2 + ax + 1$ is irreducible over F_4, but it splits over F_{16}, where it has the two roots ab and $ab + a$, where b is a root of $x^2 + x + a$ in F_{16}.

This is a special case of Artin–Schreier theory.

Imaginary Unit

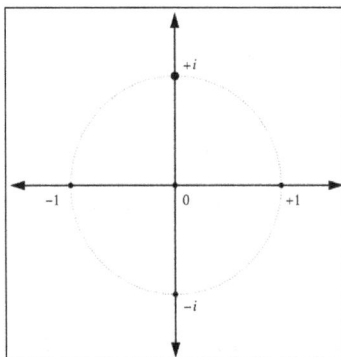

i in the complex or cartesian plane. Real numbers lie on the horizontal axis, and imaginary numbers lie on the vertical axis.

The imaginary unit or unit imaginary number (*i*) is a solution to the quadratic equation $x^2 + 1 = 0$. Although there is no real number with this property, *i* can be used to extend the real numbers to what are called complex numbers, using addition and multiplication. A simple example of the use of *i* in a complex number is $2 + 3i$.

Imaginary numbers are an important mathematical concept, which extend the real number system $\mathbb{R}$ to the complex number system $\mathbb{C}$, which in turn provides at least one root for every nonconstant polynomial $P(x)$. The term "imaginary" is used because there is no real number having a negative square.

There are two complex square roots of −1, namely i and −i, just as there are two complex square roots of every real number other than zero, which has one double square root.

In contexts where i is ambiguous or problematic, j or the Greek ι is sometimes used. In the disciplines of electrical engineering and control systems engineering, the imaginary unit is normally denoted by j instead of i, because i is commonly used to denote electric current.

The powers of i return cyclic values:
... (repeats the pattern from blue area)
$i^{-3} = i$
$i^{-2} = -1$
$i^{-1} = -i$
$i^{0} = 1$
$i^{1} = i$
$i^{2} = -1$
$i^{3} = -i$
$i^{4} = 1$
$i^{5} = i$
$i^{6} = -1$
... (repeats the pattern from the blue area)

The imaginary number i is defined solely by the property that its square is −1:

$$i^2 = -1.$$

With i defined this way, it follows directly from algebra that i and −i are both square roots of −1.

Although the construction is called "imaginary", and although the concept of an imaginary number may be intuitively more difficult to grasp than that of a real number, the construction is perfectly valid from a mathematical standpoint. Real number operations can be extended to imaginary and complex numbers by treating i as an unknown quantity while manipulating an expression, and then using the definition to replace any occurrence of i^2 with −1. Higher integral powers of i can also be replaced with −i, 1, i, or −1:

$$i^3 = i^2 i = (-1)i = -i$$
$$i^4 = i^3 i = (-i)i = -(i^2) = -(-1) = 1$$
$$i^5 = i^4 i = (1)i = i$$

Similarly, as with any non-zero real number:

$$i^0 = i^{1-1} = i^1 i^{-1} = i^1 \frac{1}{i} = i\frac{1}{i} = \frac{i}{i} = 1$$

As a complex number, i is represented in rectangular form as $0 + 1 \cdot i$, with a zero real component and a unit imaginary component. In polar form, i is represented as $1 \cdot e^{i\pi/2}$ (or just $e^{i\pi/2}$), with an absolute value (or magnitude) of 1 and an argument (or angle) of $\pi/_2$. In the complex plane (also known as the Argand plane), which is a special interpretation of a Cartesian plane, i is the point located one unit from the origin along the imaginary axis (which is orthogonal to the real axis).

i and $-i$

Being a quadratic polynomial with no multiple root, the defining equation $x^2 = -1$ has *two* distinct solutions, which are equally valid and which happen to be additive and multiplicative inverses of each other. More precisely, once a solution i of the equation has been fixed, the value $-i$, which is distinct from i, is also a solution. Since the equation is the only definition of i, it appears that the definition is ambiguous (more precisely, not well-defined). However, no ambiguity results as long as one or other of the solutions is chosen and labelled as "i", with the other one then being labelled as $-i$. This is because, although $-i$ and i are not *quantitatively* equivalent (they *are* negatives of each other), there is no *algebraic* difference between i and $-i$. Both imaginary numbers have equal claim to being the number whose square is -1. If all mathematical textbooks and published literature referring to imaginary or complex numbers were rewritten with $-i$ replacing every occurrence of $+i$ (and therefore every occurrence of $-i$ replaced by $-(-i) = +i$), all facts and theorems would continue to be equivalently valid. The distinction between the two roots x of $x^2 + 1 = 0$ with one of them labelled with a minus sign is purely a notational relic; neither root can be said to be more primary or fundamental than the other, and neither of them is "positive" or "negative".

The issue can be a subtle one. The most precise explanation is to say that although the complex field, defined as $\mathbb{R}[x]/(x^2 + 1)$, is unique up to isomorphism, it is *not* unique up to a *unique* isomorphism — there are exactly two field automorphisms of $\mathbb{R}[x]/(x^2 + 1)$ which keep each real number fixed: the identity and the automorphism sending x to $-x$.

Matrices

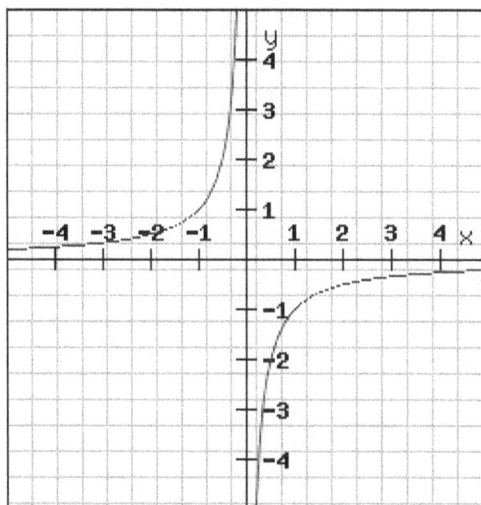

(x,y) is confined by hyperbola xy = −1 for an imaginary unit matrix.

A similar issue arises if the complex numbers are interpreted as 2 × 2 real matrices (see matrix representation of complex numbers), because then both

$$X = \begin{pmatrix} 0 & -1 \\ 1 & 0 \end{pmatrix} \quad \text{and} \quad X = \begin{pmatrix} 0 & 1 \\ -1 & 0 \end{pmatrix}$$

are solutions to the matrix equation

$$X^2 = -I = -\begin{pmatrix} 1 & 0 \\ 0 & 1 \end{pmatrix} = \begin{pmatrix} -1 & 0 \\ 0 & -1 \end{pmatrix}.$$

In this case, the ambiguity results from the geometric choice of which "direction" around the unit circle is "positive" rotation. A more precise explanation is to say that the automorphism group of the special orthogonal group SO(2, $\mathbb{R}$) has exactly two elements—the identity and the automorphism which exchanges "CW" (clockwise) and "CCW" (counter-clockwise) rotations.

All these ambiguities can be solved by adopting a more rigorous definition of complex number, and explicitly *choosing* one of the solutions to the equation to be the imaginary unit. For example, the ordered pair (0, 1), in the usual construction of the complex numbers with two-dimensional vectors.

Consider the matrix equation $\begin{pmatrix} z & x \\ y & -z \end{pmatrix}^2 = \begin{pmatrix} -1 & 0 \\ 0 & -1 \end{pmatrix}.$

Then $z^2 + xy = -1$ so the product xy is negative because $xy = -(1+z^2)$, thus the point (x, y) lies in quadrant II or IV. Furthermore,

$z^2 = -(1+xy) \geq 0 \Rightarrow xy \leq -1$ so (x,y) is bounded by the hyperbola $xy = -1$.

Proper Use

The imaginary unit is sometimes written $\sqrt{-1}$ in advanced mathematics contexts (as well as in less advanced popular texts). However, great care needs to be taken when manipulating formulas involving radicals. The radical sign notation is reserved either for the principal square root function, which is *only* defined for real $x \geq 0$, or for the principal branch of the complex square root function. Attempting to apply the calculation rules of the principal (real) square root function to manipulate the principal branch of the complex square root function can produce false results:

$$-1 = i \cdot i = \sqrt{-1} \cdot \sqrt{-1} = \sqrt{(-1) \cdot (-1)} = \sqrt{1} = 1 \ (\textit{incorrect}).$$

Similarly:

$$\frac{1}{i} = \frac{\sqrt{1}}{\sqrt{-1}} = \sqrt{\frac{1}{-1}} = \sqrt{\frac{-1}{1}} = \sqrt{-1} = i \ (\textit{incorrect}).$$

The calculation rules

$$\sqrt{a} \cdot \sqrt{b} = \sqrt{a \cdot b}$$

and

$$\frac{\sqrt{a}}{\sqrt{b}} = \sqrt{\frac{a}{b}}$$

are only valid for real, non-negative values of a and b.

These problems are avoided by writing and manipulating expressions like $i\sqrt{7}$, rather than $\sqrt{7}$.

Properties

Square Roots

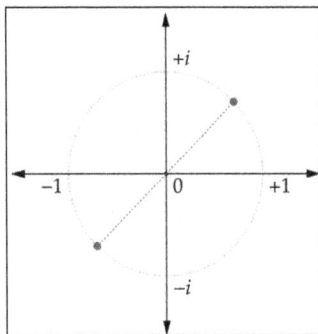

The two square roots of i in the complex plane.

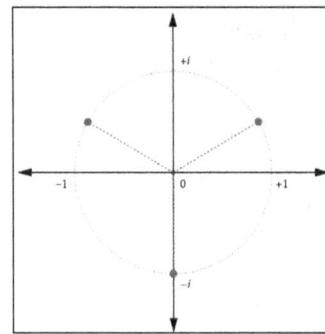

The three cube roots of i in the complex plane.

i has two square roots, just like all complex numbers (except zero, which has a double root). These two roots can be expressed as the complex numbers:

$$\pm\left(\frac{\sqrt{2}}{2} + \frac{\sqrt{2}}{2}i\right) = \pm\frac{\sqrt{2}}{2}(1+i).$$

Indeed, squaring both expressions:

$$\left(\pm\frac{\sqrt{2}}{2}(1+i)\right)^2 = \left(\pm\frac{\sqrt{2}}{2}\right)^2 (1+i)^2$$

$$= \frac{1}{2}(1 + 2i + i^2)$$

$$= \frac{1}{2}(1 + 2i - 1)$$

$$= i.$$

Using the radical sign for the principal square root gives:

$$\sqrt{i} = \frac{\sqrt{2}}{2}(1+i).$$

Cube Roots

The three cube roots of i are:

$$-i,$$

$$\frac{\sqrt{3}}{2} + \frac{i}{2},$$

$$-\frac{\sqrt{3}}{2} + \frac{i}{2}.$$

Similar to all of the roots of 1, all of the roots of i are the vertices of regular polygons inscribed within the unit circle in the complex plane.

Multiplication and Division

Multiplying a complex number by i gives:

$$i(a+bi) = ai + bi^2 = -b + ai.$$

(This is equivalent to a 90° counter-clockwise rotation of a vector about the origin in the complex plane).

Dividing by i is equivalent to multiplying by the reciprocal of i:

$$\frac{1}{i} = \frac{1}{i}\cdot\frac{i}{i} = \frac{i}{i^2} = \frac{i}{-1} = -i.$$

Using this identity to generalize division by i to all complex numbers gives:

$$\frac{a+bi}{i} = -i(a+bi) = -ai - bi^2 = b - ai.$$

(This is equivalent to a 90° clockwise rotation of a vector about the origin in the complex plane).

Powers

The powers of i repeat in a cycle expressible with the following pattern, where n is any integer:

$$i^{4n} = 1$$

$$i^{4n+1} = i$$

$$i^{4n+2} = -1$$

$$i^{4n+3} = -i,$$

This leads to the conclusion that:

$$i^n = i^{n \bmod 4}$$

where *mod* represents the modulo operation. Equivalently:

$$i^n = \cos(n\pi/2) + i\sin(n\pi/2)$$

i Raised to the Power of i

Making use of Euler's formula, i^i is

$$i^i = \left(e^{i(\pi/2+2k\pi)}\right)^i = e^{i^2(\pi/2+2k\pi)} = e^{-(\pi/2+2k\pi)}$$

where $k \in \mathbb{Z}$, the set of integers.

The principal value (for $k = 0$) is $e^{-\pi/2}$ or approximately 0.207879576...

Factorial

The factorial of the imaginary unit i is most often given in terms of the gamma function evaluated at $1 + i$:

$$i! = \Gamma(1+i) \approx 0.4980 - 0.1549i.$$

Also,

$$|i!| = \sqrt{\frac{\pi}{\sinh \pi}}$$

Other Operations

Many mathematical operations that can be carried out with real numbers can also be carried out with i, such as exponentiation, roots, logarithms, and trigonometric functions. All of the following functions are complex multi-valued functions, and it should be clearly stated which branch of the Riemann surface the function is defined on in practice. Listed below are results for the most commonly chosen branch.

A number raised to the ni power is:

$$x^{ni} = \cos(n \ln x) + i\sin(n \ln x).$$

The nith root of a number is:

$$\sqrt[ni]{x} = \cos\left(\frac{\ln x}{n}\right) - i\sin\left(\frac{\ln x}{n}\right).$$

The imaginary-base logarithm of a number is:

$$\log_i(x) = \frac{2\ln x}{i\pi}.$$

As with any complex logarithm, the log base i is not uniquely defined.

The cosine of i is a real number:

$$\cos i = \cosh 1 = \frac{e + 1/e}{2} = \frac{e^2 + 1}{2e} \approx 1.54308064\ldots$$

And the sine of i is purely imaginary:

$$\sin i = i \sinh 1 = \frac{e - 1/e}{2} i = \frac{e^2 - 1}{2e} i \approx (1.17520119\ldots)i.$$

Alternative Notations

- In electrical engineering and related fields, the imaginary unit is normally denoted by j to avoid confusion with electric current as a function of time, traditionally denoted by $i(t)$ or just i. The Python programming language also uses j to mark the imaginary part of a complex number. MATLAB associates both i and j with the imaginary unit, although $1i$ or $1j$ is preferable, for speed and improved robustness.

- Some texts use the Greek letter iota (ι) for the imaginary unit, to avoid confusion, especially with index and subscripts.

- Each of i, j, and k is an imaginary unit in the quaternions. In bivectors and biquaternions an additional imaginary unit h is used.

CUBIC EQUATION

Cubic equations and the nature of their roots.

A cubic equation has the form

$$ax^3 + bx^2 + cx + d = 0$$

It must have the term in x^3 or it would not be cubic (and so a $a \neq 0$), but any or all of b, c and d can be zero. For instance,

$$x^3 - 6x^2 + 11x - 6 = 0, \qquad 4x^3 + 57 = 0, \quad x^3 + 9x = 0$$

are all cubic equations.

Just as a quadratic equation may have two real roots, so a cubic equation has possibly three. But unlike a quadratic equation which may have no real solution, a cubic equation always has at least one real root. We will see why this is the case later. If a cubic does have three roots, two or even all three of them may be repeated. This gives us four possibilities which are illustrated in the following examples.

1. Suppose we wish to solve the equation

$$x^3 - 6x^2 + 11x - 6 = 0$$

This equation can be factorised to give

$$(x-1)(x-2)(x-3)=0$$

This equation has three real roots, all different - the solutions are $x=1$, $x=2$, and $x=3$. In figure we show the graph of $y=x^3-6x^2+11x-6$.

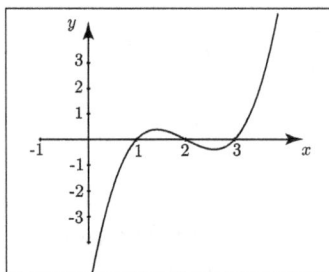

The graph of $y=x^3-6x^2+11x-6$.

Notice that it starts low down on the left, because as x gets large and negative so does x 3 and it finishes higher to the right because as x gets large and positive so does x^3 . The curve crosses the x-axis three times, once where $x = 1$, once where $x = 2$ and once where $x = 3$. This gives us our three separate solutions.

2. Suppose we wish to solve the equation $x^3-5x^2+8x-4=0$.

This equation can be factorised to give

$$(x-1)(x-2)^2=0$$

In this case we do have three real roots but two of them are the same because of the term $(x–2)^2$. So we only have two distinct solutions. Figure shows a graph of $y = x^3 – 5x^2 + 8x – 4$.

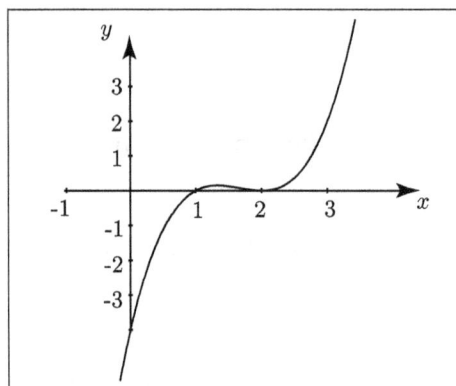

The graph of $y = x^3 – 5x^2 + 8x – 4$.

Again the curve starts low to the left and goes high to the right. It crosses the x-axis once and then just touches it at $x = 2$. So we have our two roots $x = 1$ and $x = 2$.

3. Suppose we wish to solve the equation $x^3-3x^2+3x-1=0$.

This equation can be factorised to give

$$(x-1)^3 = 0$$

So although there are three factors, they are all the same and we only have a single solution $x = 1$. The corresponding curve is $y = x^3 - 3x^2 + 3x - 1$ and is shown in figure.

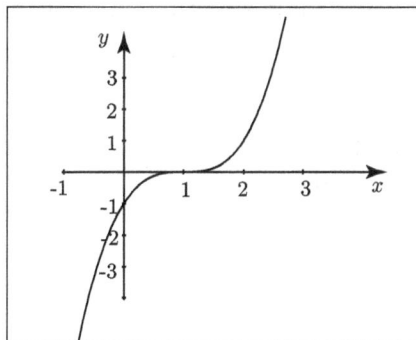

The graph of $y = x^3 - 3x^2 + 3x - 1$.

As with all the cubics we have seen so far, it starts low down on the left and goes high up to the right. Notice that the curve does cross the x-axis at the point $x = 1$ but the x-axis is also a tangent to the curve at this point. This is indicative of the fact that there are three repeated roots.

4. Suppose we wish to solve the equation $x^3 + x^2 + x - 3 = 0$.

This equation can be factorised to give

$$(x-1)(x^2 + 2x + 3) = 0$$

The quadratic $x^2 + 2x + 3 = 0$ has no real solutions, so the only solution to the cubic equation is obtained by putting $x - 1 = 0$, giving the single real solution $x = 1$.

The graph $y = x^3 + x^2 + x - 3$ is shown in figure.

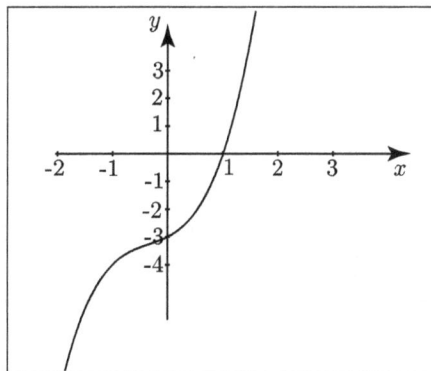

The graph of $y = x^3 + x^2 + x - 3$.

You can see that the graph crosses the x-axis in one place only.

From these graphs you can see why a cubic equation always has at least one real root. The graph either starts large and negative and finishes large and positive (when the coefficient of x^3 is positive), or it will start large and positive and finish off large and negative (when the coefficient of x^3 is negative).

The graph of a cubic must cross the x-axis at least once giving you at least one real root. So, any problem you get that involves solving a cubic equation will have a real solution.

Solving Cubic Equations

Now let us move on to the solution of cubic equations. Like a quadratic, a cubic should always be re-arranged into its standard form, in this case.

$$ax^3 + bx^2 + cx + d = 0$$

The equation

$$x^2 + 4x - 1 = \frac{6}{x}$$

is a cubic, though it is not written in the standard form. We need to multiply through by x, giving us

$$x^3 + 4x^2 - x = 6$$

This is now in the standard form

When solving cubics it helps if you know one root to start with.

1. Suppose we wish to solve

$$x^3 - 5x^2 - 2x + 24 = 0$$

given that x = −2 is a solution.

There is a theorem called the Factor Theorem which we do not prove here. It states that if $x = -2$ is a solution of this equation, then $x+2$ is a factor of this whole expression. This means that $x^3 - 5x^2 - 2x + 24 = 0$ can be written in the form

$$\left(x+2\right)\left(x^2 + ax + b\right) = 0$$

Where a and b are numbers.

Our task now is to find a and b, and we do this by a process called synthetic division. This involves looking at the coefficients of the original cubic equation, which are 1, −5, −2 and 24. These are written down in the first row of a table, the starting layout for which is

$$
\begin{array}{cccc|l}
1 & -5 & -2 & 24 & x = -2 \\
\hline
1 & & &
\end{array}
$$

Notice that to the right of the vertical line we write down the known root x = −2. We have left a blank line which will be filled in shortly. In the first position on the bottom row we have brought down the number 1 from the first row.

The next step is to multiply the number 1, just brought down, by the known root, –2, and write the result, –2, in the blank row in the position shown.

$$
\begin{array}{rrrr|l}
1 & -5 & -2 & 24 & x=-2 \\
 & -2 & & & \\
\hline
1 & & & &
\end{array}
$$

The numbers in the second column are then added, –5 + –2 = –7, and the result written in the bottom row as shown.

$$
\begin{array}{rrrr|l}
1 & -5 & -2 & 24 & x=-2 \\
 & -2 & & & \\
\hline
1 & -7 & & &
\end{array}
$$

Then, the number just written down, –7, is multiplied by the known root, –2, and we write the result, 14, in the blank row in the position shown.

$$
\begin{array}{rrrr|l}
1 & -5 & -2 & 24 & x=-2 \\
 & -2 & 14 & & \\
\hline
1 & -7 & & &
\end{array}
$$

Then the numbers in this column are added:

$$
\begin{array}{rrrr|l}
1 & -5 & -2 & 24 & x=-2 \\
 & -2 & 14 & & \\
\hline
1 & -7 & 12 & &
\end{array}
$$

The process continues:

$$
\begin{array}{rrrr|l}
1 & -5 & -2 & 24 & x=-2 \\
 & -2 & 14 & -24 & \\
\hline
1 & -7 & 12 & 0 &
\end{array}
$$

Note that the final number in the bottom row (obtained by adding 24 and –24) is zero. This is confirmation that $x = -2$ is a root of the original cubic. If this value turns out to be non-zero then we do not have a root.

At this stage the coefficients in the quadratic that we are looking for are the first three numbers in the bottom row. So the quadratic is

$$x^2 - 7x + 12$$

So we have reduced our cubic to

$$(x+2)(x^2 - 7x + 12) = 0$$

The quadratic term can be factorised to give

$$(x + 2)(x - 3)(x - 4) = 0$$

giving us the solutions $x = -2,\ 3$ or 4

In the previous Example we were given one of the roots. If a root is not known it's always worth trying a few simple values.

2. Suppose we wish to solve

$$x^3 - 7x - 6 = 0$$

A very simple value we might try is $x = 1$. Substituting $x = 1$ in the left-hand side we find

$$1^3 - 7(1) - 6$$

Which equals -12, so this value is clearly not a solution. Suppose we try the value $x = -1$:

$$(-1)^3 - 7(-1) - 6$$

which does equal zero, so $x = -1$ is a solution. This means that $x + 1$ is a factor and the cubic can be written in the form

$$(x+1)(x^2 + ax + b) = 0$$

We can perform synthetic division to find the other factors.

As before, we take the coefficients of the original cubic equation, which are 1, 0, −7 and −6. These are written down in the first row of a table, the starting layout for which is

$$
\begin{array}{cccc|l}
1 & 0 & -7 & -6 & x = -1 \\
\hline
1 & & & &
\end{array}
$$

To the right of the vertical line we write down the known root $x = -1$. In the first position on the bottom row we have brought down the number 1 from the first row.

The next step is to multiply the number 1, just brought down, by the known root, −1, and write the result, −1, in the blank row in the position shown.

$$
\begin{array}{cccc|l}
1 & 0 & -7 & -6 & x = -1 \\
 & -1 & & & \\
\hline
1 & & & &
\end{array}
$$

To the right of the vertical line we write down the known root $x = -1$. In the first position on the bottom row we have brought down the number 1 from the first row.

The numbers in the second column are then added, $0 + -1 = -1$, and the result written in the bottom row as shown.

$$\begin{array}{rrrr|l} 1 & 0 & -7 & -6 & x = -1 \\ & -1 & & & \\ \hline 1 & -1 & & & \end{array}$$

Then, the number just written down, −1, is multiplied by the known root, −1, and we write the result, 1, in the blank row in the position shown.

$$\begin{array}{rrrr|l} 1 & 0 & -7 & -6 & x = -1 \\ & -1 & & & \\ \hline 1 & -1 & & & \end{array}$$

Then the numbers in this column are added:

$$\begin{array}{rrrr|l} 1 & 0 & -7 & -6 & x = -1 \\ & -1 & 1 & & \\ \hline 1 & -1 & -6 & & \end{array}$$

The process continues:

$$\begin{array}{rrrr|l} 1 & 0 & -7 & -6 & x = -1 \\ & -1 & 1 & 6 & \\ \hline 1 & -1 & -6 & 0 & \end{array}$$

At this stage the coefficients in the quadratic that we are looking for are the first three numbers in the bottom row. So the quadratic is

$$x^2 - x - 6$$

So we now need to solve the equation

$$(x+1)(x^2 - x - 6) = 0$$

which factorises to give

$$(x+1)(x-3)(x+2) = 0$$

and the three solutions to the cubic equation are $x = -2, -1 \text{ or } 3$.

Sometimes you may be able to spot a factor as shown in the following Example.

3. Suppose we wish to solve the equation $x^3 - 4x^2 - 9x + 36 = 0$.

Observe that the first two terms in the cubic can be factorised as $x^3 - 4x^2 = x^2(x - 4)$. The second pair of terms can be factorised as $-9x + 36 = -9(x - 4)$. This sort of observation can only be made when you have had sufficient practice and experience of handling expressions like this. However, the observation enables us to proceed as follows:

$$x^3 - 4x^2 - 9x + 36 \ \ = 0$$
$$x^2(x-4) - 9(x-4) = 0$$

The common factor of $(x - 4)$ can be extracted to give

$$(x^2 - 9)(x - 4) - 0$$

and the difference of two squares can be factorised giving

$$(x + 3)(x - 3)(x - 4) = 0$$

giving us solutions $x = -3, \ 3$ or 4

You may have noticed that in each example we have done, every root is a factor of the constant term in the equation. In the last Example, -3, 3 and 4 all divide into the constant term 36. As long as the coefficient of x^3 in the cubic equation is 1 this must be the case. This is because, referring to equation $(x+3)(x-3)(x-4) = 0$ for example, the constant term arises from multiplying the numbers 3, -3 and -4. This gives us another possible approach.

4. Suppose we wish to solve the equation

$$x^3 - 6x^2 - 6x - 7 = 0$$

If there is to be a solution, then because the coefficient of x^3 is 1, it is going to be an integer and it's going to be a factor of 7. This leaves us with only four possibilities: 1, -1, 7, and -7. So we can try each of them in turn. You can see fairly quickly that 1 and -1 don't work. So we will try 7. Rather than substitute $x = 7$ into the cubic equation we shall immediately try to synthetically divide this expression by 7.

$$
\begin{array}{rrrr|l}
1 & -6 & -6 & -7 & x = 7 \\
 & 7 & 7 & 7 & \\
\hline
1 & 1 & 1 & 0 &
\end{array}
$$

Because the fourth number in the last row is zero it follows that $x = 7$ is indeed a root. Had this number not been zero then $x = 7$ would not have been a root. So 7 is indeed a root, and the resulting quadratic is

$$x^2 + x + 1 = 0$$

Now you will find that if you try to solve it that the quadratic equation $x^2 + x + 1 = 0$ has no real solutions, so the only possible solution to this cubic is $x = 7$.

Certain basic identities which you may wish to learn can help in factorising both cubic and quadratic equations.

5. Suppose we wanted to solve the equation $x^3 + 3x^2 + 3x + 1 = 0$.

This has the widely-known factorisation $(x + 1)^3 = 0$ from which we have the root $x = -1$ repeated three times.

Using Graphs to Solve Cubic Equations

If you cannot find a solution by these methods then draw an accurate graph of the cubic expression. The points where it crosses the x-axis will give you the solutions to the equation but their accuracy will be limited to the accuracy of your graph.

You may indeed find that the graph crosses the x-axis at a point that would suggest a factor. for instance, if you draw a graph that appears to cross the x-axis at, say, $x = \dfrac{1}{2}$ then it's worth trying to find out whether $\left(x = \dfrac{1}{2} \right)$ is indeed a factor.

1. Suppose we wished to solve $x^3 + 4x^2 + x - 5 = 0$.

Now this equation will not yield a factor by any of the methods that we have discussed. So a graph of $y = x^3 + 4x^2 + x - 5$ has been drawn as shown in figure.

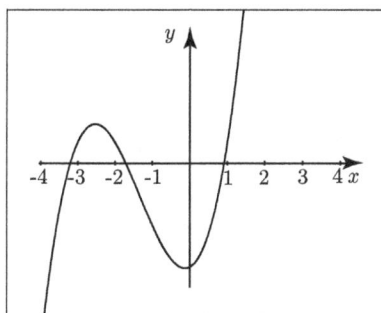

A graph of $y = x^3 + 4x^2 + x - 5$.

It crosses the x-axis at three places and hence there are three real roots. As we stated before their accuracy will be limited to the accuracy of the graph. From the graph we find the approximate solutions $x \approx -3.2, -1.7, 0.9$.

QUARTIC EQUATION

A quartic equation is a fourth-order polynomial equation of the form

$$z^4 + a_3 z^3 + a_2 z^2 + a_1 z + a_0 = 0.$$

While some authors use the term "biquadratic equation" as a synonym for quartic equation, others reserve the term for a quartic equation having no cubic term, i.e., a quadratic equation in x^2.

Ferrari was the first to develop an algebraic technique for solving the general quartic, which was stolen and published in Cardano's *Ars Magna* in 1545. The Wolfram Language can solve quartic equations exactly using the built-in command Solve $[a^4 x^4 + a^3 x^3 + a^2 x^2 + ax + a\theta == 0, x]$.

The solution can also be expressed in terms of Wolfram Language algebraic root objects by first issuing Set Options[Roots, Quartics -> False].

The roots of this equation satisfy Vieta's formulas:

$$x_1 + x_2 + x_3 + x_4 = -a_3$$
$$x_1 x_2 + x_1 x_3 + x_1 x_4 + x_2 x_3 + x_2 x_4 + x_3 x_4 = a_2$$
$$x_1 x_2 x_3 + x_2 x_3 x_4 + x_1 x_2 x_4 + x_1 x_3 x_4 = -a_1$$
$$x_1 x_2 x_3 x_4 = a_0,$$

where the denominators on the right side are all $a_4 \equiv 1$. Writing the quartic in the standard form

$$x^4 + px^2 + qx + r = 0,$$

the properties of the symmetric polynomials appearing in Vieta's formulas then give

$$z_1^2 + z_2^2 + z_3^2 + z_4^2 = -2p$$
$$z_1^3 + z_2^3 + z_3^3 + z_4^3 = -3q$$
$$z_1^4 + z_2^4 + z_3^4 + z_4^4 = 2p^2 - 4r$$
$$z_1^5 + z_2^5 + z_3^5 + z_4^5 = -5pq.$$

Eliminating p, q, and r, respectively, gives the relations

$$z_1 z_2 \left(p + z_1^2 + z_1 z_2 + z_2^2\right) - r = 0$$
$$z_1^2 z_2 \left(z_1 + z_2\right) - q z_1 - r = 0$$
$$q + pz_2 + z_2^3 = 0,$$

as well as their cyclic permutations.

Ferrari was the first to develop an algebraic technique for solving the general quartic. He applied his technique (which was stolen and published by Cardano) to the equation.

$$x^4 + 6x^2 - 60x + 36 = 0$$

The x^3 term can be eliminated from the general quartic ($\Diamond$) by making a substitution of the form

$$z \equiv x - \lambda,$$

so

$$x^4 + \left(a_3 - 4\lambda\right)x^3 + \left(a_2 - 3a_3 \lambda + 6\lambda^2\right)x^2$$
$$+ \left(a_1 - 2a_2\lambda + 3a_3 \lambda^2 - 4\lambda^3\right)x + \left(a_o - a_1 \lambda + a_2 \lambda^2 - a_3 \lambda^3 + \lambda^4\right).$$

Letting $\lambda = a_3/4$ so

$$z \equiv x - \frac{1}{4}a_3$$

then gives the standard form

$$x^4 + px^2 + qx + r = 0,$$

where

$$p \equiv a_2 - \frac{3}{8}a_3^2$$

$$q \equiv a_1 - \frac{1}{2}a_2 a_3 + \frac{1}{8}a_3^3$$

$$r \equiv a_0 - \frac{1}{4}a_1 a_3 + \frac{1}{16}a_2 a_3^2 - \frac{3}{256}a_3^4.$$

The quartic can be solved by writing it in a general form that would allow it to be algebraically factorable and then finding the condition to put it in this form. The equation that must be solved to make it factorable is called the resolvent cubic. To do this, note that the quartic will be factorable if it can be written as the difference of two squared terms,

$$P^2 - Q^2 = (P+Q)(P-Q).$$

It turns out that a factorization of this form can be obtained by adding and subtracting $x^2 u + u^2/4$ (where u is for now an arbitrary quantity, but which will be specified shortly) to equation ($\diamond$) to obtain

$$\left(x^4 + x^2 u + \frac{1}{4}u^2\right) - x^2 u - \frac{1}{4}u^2 + px^2 + qx + r = 0.$$

This equation can be rewritten:

$$\left(x^2 + \frac{1}{2}u\right)^2 - \left[(u-p)x^2 - qx + \left(\frac{1}{4}u^2 - r\right)\right] = 0.$$

Note that the first term is immediately a perfect square P^2 with

$$P \equiv x^2 + \frac{1}{2}u,$$

and the second term will be a perfect square Q^2 if u is chosen to that the square can be completed in

$$Q^2 = (u-p)\left(x^2 - \frac{q}{u-p}x + \frac{\frac{1}{4}u^2 - r}{u-p}\right).$$

This means we want

$$Q^2 = (u-p)\left(x - \sqrt{\frac{\frac{1}{4}u^2 - r}{u-p}}\right)^2$$

which requires that

$$2\sqrt{\frac{\frac{1}{4}u^2-r}{u-p}}=\frac{q}{u-p},$$

or

$$q^2=4(u-p)\left(\frac{1}{4}u^2-r\right).$$

This is the resolvent cubic.

Since an analytic solution to the cubic is known, we can immediately solve algebraically for one of the three solution of equation $q^2=4(u-p)\left(\frac{1}{4}u^2-r\right)$, say u_1, and plugging equation

$$q^2=(u-p)\left(-u^2-r\right) \text{ into equation } Q^2=(u-p)\left(x^2-\frac{q}{u-p}x+\frac{\frac{1}{4}u^2-r}{u-p}\right). \text{ then gives}$$

$$Q=Ax-\frac{q}{2A}$$

with

$$A\equiv\sqrt{u_1-p}.$$

Q therefore is linear in x and P is quadratic in x, so each term $P+Q$ and $P-Q$ is quadratic and can be solved using the quadratic formula, thus giving all four solutions to the original quartic.

Explicitly, plugging p, q, and r back into ($\diamondsuit$) gives

$$u^3+\left(\frac{3}{8}a_3^2-a_2\right)u^2+\left(\frac{3}{64}a_3^4-\frac{1}{4}a_2\,a_3^2+a_1\,a_3-4\,a_0\right)u+$$

$$\left(\frac{1}{512}a_3^6-\frac{1}{64}a_2\,a_3^4+\frac{1}{8}a_1\,a_3^3-\frac{3}{2}a_0\,a_3^2+4\,a_0\,a_2-a_1^2\right).$$

This can be simplified by making the substitution

$$u=y-\frac{1}{8}a_3^2,$$

which gives the resolvent cubic equation

$$y^3-a_2\,y^2+(a_1\,a_3-4\,a_0)y+(4\,a_2\,a_0-a_1^2-a_3^2\,a_0)=0.$$

Let y_1 be a real root of (34), then the four roots of the original quartic are given by the roots of the equation

$$x^2 + \frac{1}{2}\left(a_3 \pm \sqrt{a_3^2 - 4a_2 + 4y_1}\right)x + \frac{1}{2}\left(y_1 \pm \sqrt{y_1^2 - 4a_0}\right) = 0,$$

which are

$$z_1 = -\frac{1}{4}a_3 + \frac{1}{2}R + \frac{1}{2}D$$

$$z_2 = -\frac{1}{4}a_3 + \frac{1}{2}R - \frac{1}{2}D$$

$$z_3 = -\frac{1}{4}a_3 - \frac{1}{2}R + \frac{1}{2}E$$

$$z_4 = -\frac{1}{4}a_3 - \frac{1}{2}R - \frac{1}{2}E,$$

where

$$R \equiv \sqrt{\frac{1}{4}a_3^2 - a_2 + y_1}$$

$$D \equiv \begin{cases} \sqrt{\frac{3}{4}a_3^2 - R^2 - 2a_2 + \frac{1}{4}\left(4a_3\,a_2 - 8a_1 - a_3^3\right)R^{-1}} & \text{for } R \neq 0 \\ \\ \sqrt{\frac{3}{4}a_3^2 - 2a_2 + 2\sqrt{y_1^2 - 4a_0}} & \text{for } R \neq 0 \end{cases}$$

$$E \equiv \begin{cases} \sqrt{\frac{3}{4}a_3^2 - R^2 - 2a_2 - \frac{1}{4}\left(4a_3\,a_2 - 8a_1 - a_3^3\right)R^{-1}} & \text{for } R \neq 0 \\ \\ \sqrt{\frac{3}{4}a_3^2 - 2a_2 - 2\sqrt{y_1^2 - 4a_0}} & \text{for } R \neq 0 \end{cases}$$

Another approach to solving the quartic ($\diamond$) defines

$$\alpha \equiv (x_1 + x_2)(x_3 + x_4) = -(x_1 + x_2)^2$$

$$\beta \equiv (x_1 + x_3)(x_2 + x_4) = -(x_1 + x_3)^2$$

$$\gamma \equiv (x_1 + x_4)(x_2 + x_3) = -(x_2 + x_3)^2,$$

where the second forms follow from

$$x_1 + x_2 + x_3 + x_4 = -a_3 = 0,$$

and defining

$$h(x) \equiv (x - \alpha) + (x - \beta)(x - \gamma)$$

$$= x^3 - (\alpha + \beta + \gamma)x^2 + (\alpha\beta + \alpha\gamma + \beta\gamma)x - \alpha\beta\gamma.$$

This equation can be written in terms of the original coefficients , p q ,and r as

$$h(x) = x^3 - 2p\,x^2 + \left(p^2 - 4r\right)x + q^2.$$

The roots of this cubic equation then give α, β, and y, and the equations ($\Diamond$) to ($\Diamond$) can be solved for the four roots x_1 of the original quartic.

QUINTIC EQUATION

Unlike quadratic, cubic, and quartic polynomials, the *general* quintic cannot be solved algebraically in terms of a finite number of additions, subtractions, multiplications, divisions, and root extractions, as rigorously demonstrated by Abel (Abel's impossibility theorem) and Galois. However, certain classes of quintic equations *can* be solved in this manner.

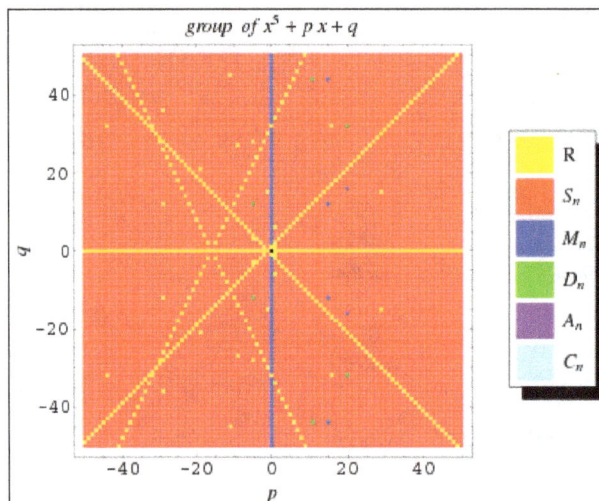

group of $x^5 + px + q$

Irreducible quintic equations can be associated with a Galois group, which may be a symmetric group S_n, metacyclic group M_n, dihedral group D_n, alternating group A_n, or cyclic group C_n, as illustrated above. Solvability of a quintic is then predicated by its corresponding group being a solvable group. An example of a quintic equation with solvable cyclic group is

$$1024x^5 - 2816x^4 + 2816x^3 - 1232x^2 + 220x - 11 = 0,$$

which arises in the computation of $\sin(\pi/11)$.

In the case of a solvable quintic, the roots can be found using the formulas found in 1771 by Malfatti, who was the first to "solve" the quintic using a resolvent of sixth degree.

The general quintic can be solved in terms of Jacobi theta functions, as was first done by Hermite in 1858. Kronecker subsequently obtained the same solution more simply, and Brioschi also derived the equation. To do so, reduce the general quintic

$$a_5 x^5 + a_4 x^4 + a_3 x^3 + a_2 x^2 + a_1 x + a_0 = 0$$

into Bring quintic form

$$x^5 - x + \rho = 0.$$

Defining,

$$k \equiv \tan\left[\frac{1}{4}\sin^{-1}\left(\frac{16}{25\sqrt{5}\,\rho^2}\right)\right]$$

$$s \equiv \begin{cases} -\mathrm{sgn}\left(\mathrm{I}[\rho]\right) & \text{for } \mathrm{R}[\rho]=0 \\ \mathrm{sgn}\left(\mathrm{R}[\rho]\right) & \text{for } \mathrm{R}[\rho]\neq 0 \end{cases}$$

$$b = \frac{s\left(k^2\right)^{1/8}}{2\cdot 5^{3/4}\sqrt{k\left(1-k^2\right)}},$$

where k is the elliptic modulus, the roots of the original quintic are then given by

$$(-1)^{3/4}\,b\left\{\left[m\left(e^{-2\pi i/5}q^{1/5}\right)\right]^{1/8}+i\left[m\left(e^{2\pi i/5}q^{1/5}\right)\right]^{1/8}\right\}$$

$$x_1 = \left\{[m]\left(e^{-4\pi i/5}q^{1/5}\right)^{1/8}+\left[m\left(e^{4\pi i/5}q^{1/5}\right)\right]^{1/8}\right\}\left\{\left[m\left(q^{1/5}\right)\right]^{1/8}+q^{5/8}\left(q^5\right)^{-1/8}\left[m\left(q^5\right)\right]^{1/8}\right\}$$

$$b\left\{-\left[m\left(q^{1/5}\right)\right]^{1/8}+e^{3\pi i/4}\left[m\left(e^{2\pi i/5}q^{1/5}\right)\right]^{1/8}\right\}\times$$

$$x_2 = \left\{e^{-3\pi i/4}\left[m\left(e^{-2\pi i/5}q^{1/5}\right)\right]^{1/8}+i\left[m\left(e^{4\pi i/5}q^{1/5}\right)\right]^{1/8}\right\}$$

$$\left\{i\left[m\left(e^{-4\pi i/5}q^{1/5}\right)\right]^{1/8}+q^{5/8}\left(q^5\right)^{-1/8}\left[m\left(q^5\right)\right]^{1/8}\right\}$$

$$b\left\{e^{-3\pi i/4}\left[m\left(e^{-2\pi i/5}\right)q^{1/5}\right]^{1/8}-i\left[m\left(e^{-4\pi i/5}q^{1/5}\right)^{1/8}\right]\right\}\times$$

$$x_3 = \left\{-\left[m\left(q^{1/5}\right)\right]^{1/8}-i\left[m\left(e^{4\pi i/5}q^{1/5}\right)\right]^{1/8}\right\}\times$$

$$\left\{e^{3\pi i/4}\left[m\left(e^{2\pi i/5}q^{1/5}\right)\right]^{1/8}+q^{5/8}\left(q^5\right)^{-1/8}\left[m\left(q^5\right)\right]^{1/8}\right\}$$

$$b\left\{\left[m\left(q^{1/5}\right)\right]^{1/8}-i\left[m\left(e^{-4\pi i/5}q^{1/5}\right)\right]^{1/8}\right\}\times$$

$$x_4 = \left\{-e^{3\pi i/5}q^{1/5}\left[m\left(e^{2\pi i/5}q^{1/5}\right)\right]^{1/8}-i\left[m\left(e^{4\pi i/5}q^{1/5}\right)\right]^{1/8}\right\}$$

$$\left\{e^{-3\pi i/4}\left[m\left(e^{-2\pi i/5}q^{1/5}\right)\right]^{1/8}+q^{5/8}\left(q^5\right)^{-1/8}\left[m\left(q^5\right)\right]^{1/8}\right\}$$

$$b\left\{\left[m\left(q^{1/5}\right)\right]^{1/8}-e^{-3\pi i/4}\left[m\left(e^{-2\pi i/5}q^{1/5}\right)\right]^{1/8}\right\}\times$$

$$x_5 = \left\{-e^{3\pi i/4}\left[m\left(e^{2\pi i/5}q^{1/5}\right)\right]^{1/8}+i\left[m\left(e^{-4\pi i/5}q^{1/5}\right)\right]^{1/8}\right\}$$

$$\left\{-i\left[m\left(e^{4\pi i/5}q^{1/5}\right)\right]^{1/8}+q^{5/8}\left(q^5\right)^{-1/8}\left[m\left(q^5\right)\right]^{1/8}\right\}.$$

where

$$m(q) = \frac{\vartheta_2^4(0,q)}{\vartheta_3^4(0,q)}$$

is the inverse nome, which is expressible as a ratio of Jacobi theta functions.

Euler reduced the general quintic to

$$x^5 - 10qx^2 - p = 0.$$

A quintic also can be algebraically reduced to principal quintic form

$$x^5 - a_2 x^2 + a_1 x + a_0 = 0.$$

By solving a quartic, a quintic can be algebraically reduced to the Bring quintic form, as was first done by Jerrard. Runge and Cadenhad and Young found a parameterization of solvable quintics in the form

$$x^5 - ax + b = 0.$$

by showing that all irreducible solvable quintics with coefficients of x^4, x^3, and x^2 missing have the following form

$$x^5 + \frac{5\mu^4(4v+3)}{v^2+1}x + \frac{4\mu^5(2v+1)(4v+3)}{v^2+1} = 0,$$

where μ and v are rational.

Spearman and Williams showed that an irreducible quintic of the form $x^5 - ax + b = 0$ having rational coefficients is solvable by radicals iff there exist rational numbers $\epsilon = \pm 1, c \geq 0$, and $e \neq 0$ such that

$$a = \frac{5e^4(3-4\epsilon c)}{c^2+1}$$

$$b = \frac{-4e^5(11\epsilon+2c)}{c^2+1}$$

The roots are then

$$x_j = e\left(\omega^j u_1 + \omega^{2j} u_2 + \omega^{3j} u_3 + \omega^{4j} u_4\right),$$

where

$$u_1 = \left(\frac{v_1^2 v_3}{D^2}\right)^{1/5}$$

$$u_2 = \left(\frac{v_3^2 v_4}{D^2}\right)^{1/5}$$

$$u_3 = \left(\frac{v_2^2 v_1}{D^2}\right)^{1/5}$$

$$u_4 = \left(\frac{v_4^2 v_2}{D^2}\right)^{1/5}$$

$$v_1 = \sqrt{D} + \sqrt{D - \epsilon \sqrt{D}}$$

$$v_2 = -\sqrt{D} - \sqrt{D + \epsilon \sqrt{D}}$$

$$v_3 = -\sqrt{D} + \sqrt{D + \epsilon \sqrt{D}}$$

$$v_4 = \sqrt{D} - \sqrt{D - \epsilon \sqrt{D}}$$

$$D = c^2 + 1.$$

Felix Klein used a Tschirnhausen transformation to reduce the general quintic to the form

$$y^5 + 5a\,y^2 + 5b\,y + c = 0.$$

He then solved the related icosahedral equation

$$I(z, I, Z) = z^5 \left(-1 + 11z^5 + z^{10}\right)^5$$
$$-\left[1 + z^{30} - 10005\left(z^{10} + z^{20}\right) + 522\left(-z^5 + z^{25}\right)\right]^2 Z = 0,$$

where Z is a function of radicals of a, b, and c. The solution of this equation can be given in terms of hypergeometric functions as

$$\frac{Z^{-1/60} \,_2F_1\left(-\frac{1}{60}, \frac{29}{60}, \frac{4}{5}, 1728Z\right)}{Z^{11/60} \,_2F_1\left(\frac{11}{60}, \frac{41}{60}, \frac{6}{5}, 1728Z\right)}.$$

Another possible approach uses a series expansion, which gives one root of the Bring quintic form. All five roots can be derived using differential equations. Let

$$F_1(\rho) = F_2(\rho)$$

$$F_2(\rho) = \,_4F_3\left(\frac{1}{5}, \frac{2}{5}, \frac{3}{5}, \frac{4}{5}; \frac{1}{2}, \frac{3}{4}, \frac{5}{4}; \frac{3125}{256}, \rho^4\right)$$

$$F_3(\rho) = \,_4F_3\left(\frac{9}{20}, \frac{13}{20}, \frac{17}{20}, \frac{21}{20}; \frac{3}{4}, \frac{5}{4}, \frac{3}{2}; \frac{3125}{256}, \rho^4\right)$$

$$F_4(\rho) = \,_4F_3\left(\frac{7}{10}, \frac{9}{10}, \frac{11}{10}, \frac{13}{10}; \frac{5}{4}, \frac{3}{2}, \frac{7}{4}; \frac{3125}{256}, \rho^4\right),$$

then the roots are

$$t_1 = -\rho_4 \, F_3\left(\frac{1}{5}, \frac{2}{5}, \frac{3}{5}, \frac{4}{5}, \frac{1}{2}, \frac{3}{4}, \frac{5}{4}, \frac{3125}{256}, \rho^4\right)$$

$$t_2 = -F_1(\rho) + \frac{1}{4}\rho F_2(\rho) + \frac{5}{32}\rho^2 F_3(\rho) + \frac{5}{32}\rho^3 F_4(\rho)$$

$$t_3 = -F_1(\rho) + \frac{1}{4}\rho F_2(\rho) - \frac{5}{32}\rho^2 F_3(\rho) + \frac{5}{32}\rho^3 F_4(\rho)$$

$$t_4 = -i\, F_1(\rho) + \frac{1}{4}\rho F_2(\rho) - \frac{5}{32}i\rho^2 F_3(\rho) - \frac{5}{32}\rho^3 F_4(\rho)$$

$$t_5 = i\, F_1(\rho) + \frac{1}{4}\rho F_2(\rho) + \frac{5}{32}i\rho^2 F_3(\rho) - \frac{5}{32}\rho^3 F_4(\rho).$$

This technique gives closed form solutions in terms of hypergeometric functions in one variable for any polynomial equation which can be written in the form

$$x^p + bx^q + c.$$

Consider the quantic

$$\prod_{j=0}^{4}\left[x - \left(\omega^j u_1 + \omega^{4j} u_2\right)\right] = 0,$$

where $\omega = e^{2\pi i/5}$ and u_1 and u_2 are complex numbers, which is related to de Moivre's quintic, and generalize it to

$$\prod_{j=0}^{4}\left[x - \left(\omega^j u_1 + \omega^{2j} u_2 + \omega^{3j} u_3 + \omega^{4j} u_4\right)\right] = 0.$$

Expanding,

$$\left(\omega^j u_1 + \omega^{2j} u_2 + \omega^{3j} u_3 + \omega^{4j} u_4\right)^5 -$$
$$5U\left(\omega^j u_1 + \omega^{2j} u_2 + \omega^{3j} u_3 + \omega^{4j} u_4\right)^3 - 5V\left(\omega^j u_1 + \omega^{2j} u_2 + \omega^{3j} u_3 + \omega^{4j} u_4\right)^2 +$$
$$5W\left(\omega^j u_1 + \omega^{2j} u_2 + \omega^{3j} u_3 + \omega^{4j} u_4\right) + \left[5(X - Y) - Z\right] = 0,$$

Where

$$U = u_1 u_4 + u_2 u_3$$

$$V = u_1 u_2^2 + u_2 u_4^2 + u_3 u_1^2 + u_4 u_3^2$$

$$W = u_1^2 u_4^2 + u_2^2 u_3^2 - u_1^3 u_2 - u_2^3 u_4 - u_3^3 u_1 - u_4^3 u_3 - u_1 u_2 u_3 u_4$$

$$X = u_1^3 u_3 u_4 + u_2^3 u_1 u_3 + u_3^3 u_2 u_4 + u_4^3 u_1 u_2$$

$$Y = u_1 u_3^2 u_4^2 + u_2 u_1^2 u_3^2 + u_3 u_2^2 u_4^2 + u_4 u_1^2 u_2^2$$

$$Z = u_1^5 + u_2^5 + u_3^5 + u_4^5$$

The u_i s satisfy

$$u_1 u_4 + u_2 u_3 = 0$$

$$u_1 u_2^2 + u_2 u_4^2 + u_3 u_1^2 + u_4 u_3^2 = 0$$

$$u_1^2 u_4^2 + u_2^2 u_3^2 - u_1^3 u_2 - u_2^3 u_4 - u_3^3 u_1 - u_4^3 u_3 - u_1 u_2 u_3 u_4$$

$$= \frac{1}{5} a$$

$$5\left[\left(u_1^3 u_3 u_4 + u_2^3 u_1 u_3 + u_3^3 u_3 u_4 + u_4^3 u_1 u_2 \right) - \left(u_1 u_3^2 u_4^2 + u_2 u_1^2 u_3^2 + u_3 u_2^2 u_4^2 + u_4 u_1^2 u_2^2 \right) \right]$$

$$- \left(u_1^5 u_2^5 + u_3^5 u_4^5 \right) = b$$

References

- Elementary-algebra, science: britannica.com, Retrieved 11 April, 2019

- James Stewart (2012), Calculus: Early Transcendentals, edition 7E, Brooks/Cole. ISBN 978-0-538-49790-9

- Mc-ty-cubicequations-2009-1, resources: mathcentre.ac.uk, Retrieved 15 March, 2019

- Hughes-Hallett, Deborah; Connally, Eric; McCallum, William G. (2007), College Algebra, John Wiley & Sons Inc., p. 205, ISBN 9780471271758

- Quartic Equation: mathworld.wolfram.com, Retrieved 20 March, 2019

Algebraic Number Theory

Algebraic number theory studies the integers, rational numbers and their generalizations with the use of abstract algebra. Class field theory and abstract analytic number theory are some of the theories that fall under its domain. This chapter discusses in detail these theories related to algebra.

ALGEBRAIC NUMBER

An algebraic number is any real number that is a solution of some single-variable polynomial equation whose coefficient s are all integer s.

A complex (sometimes, real) number that is a root of a polynomial

$$f(x) = a_n x^n + \ldots + a_1 x + a_0$$

with rational coefficients, not all of which are zero. If α is an algebraic number, then, among all polynomials with rational coefficients and $\alpha\alpha$ as a root, there exists a unique polynomial $\phi(x)$ of lowest degree with leading coefficient equal to one, which is therefore irreducible. It is called the irreducible, or minimal, polynomial of the algebraic number α. The degree n of the minimal polynomial $\phi(x)$ is also called the degree of the algebraic number α. The existence of irreducible polynomials of any degree n implies the existence of algebraic numbers of degree n. All rational numbers, and only such numbers, are algebraic numbers of the first degree. The number ii is an algebraic number of the second degree, since it is a root of the polynomial $x^2 + 1$, while $2^{1/n}$, where n is any positive integer, is an algebraic number of degree n, being a root of the irreducible polynomial $x^n - 2$.

The roots $\alpha_1, \ldots, \alpha_n$ an of the irreducible polynomial are called the conjugates of α, and are also algebraic numbers of degree n. All numbers conjugate with α are distinct. Apart from its degree, another important characteristic of an algebraic number is its height, which is the analogue of the denominator of a rational fraction. The height of an algebraic number α is the greatest absolute value of the coefficients of the irreducible and primitive polynomial with integral rational coefficients that has $\alpha\alpha$ as a root (cf. Primitive polynomial). The sum, the difference, the product and the quotient of two algebraic numbers (except for division by zero) are algebraic numbers; this means that the set of all algebraic numbers is a field. A root of a polynomial with algebraic coefficients is an algebraic number.

It was shown in 1872 by G. Cantor that the set of all algebraic numbers is denumerable, and this implied the existence of transcendental numbers (cf. Transcendental number).

An algebraic number is called an algebraic integer if all the coefficients of its minimal polynomial are rational integers. For instance, i and $1 + \sqrt{2}$ are algebraic integers, being roots of the polynomials $x^2 + 1$ and $x^2 - 2x - 1$.

The concept of an algebraic integer is a generalization of the concept of a rational integer (a rational integer mm is an algebraic integer, being the root of the polynomial $x - m$). Many properties of rational integers are also displayed by algebraic integers. Thus, the algebraic integers form a ring; on the other hand, the real algebraic integers form an everywhere-dense set in R, while the rational integers form a discrete set.

A root of any (not necessarily irreducible) polynomial with rational integer coefficients and leading coefficient one is an algebraic integer. Moreover, a root of a polynomial with algebraic integer coefficients and with leading coefficient one is an algebraic integer. In particular the k-th degree root of an algebraic integer is an algebraic integer. For any algebraic number α there exists a positive integer r such that $r\alpha$ is an algebraic integer (in analogy with rational numbers). The smallest possible number r is the modulus of the leading coefficient of the irreducible primitive polynomial with rational integer coefficients that has α as a root. All conjugates of an algebraic integer are also algebraic integers.

One says that an algebraic integer β is divisible by an algebraic integer α, $\alpha \neq 0$, if there exists an algebraic integer γ such that $\beta = \gamma\,\alpha$ Many divisibility characteristics of rational integers also hold for algebraic integers.

An algebraic integer $\in$ is called an algebraic unit (or unit for short) if it is a divisor of 1, that is, if $1/\in$ is an algebraic integer. A unit is a divisor of any algebraic integer. The inverse of a unit is a unit; the numbers conjugate to a unit are units; all divisors of a unit are units; a product of a finite number of units is a unit. An algebraic integer is a unit if and only if the product of all its conjugates is ±1. The k-th roots of unity are units, each one having modulus 1. There exists an infinite set of other units which do not have modulus 1. For instance, the numbers $2 - \sqrt{3}$ and $2 + \sqrt{3}$ are units, being the roots of the polynomial $x^2 - 4x + 1$. Moreover, their powers constitute units which may be arbitrary large or small. The field of rational numbers contains only two units — +1 and −1.

Two algebraic integers are called associated if they differ by a factor which is a unit. There is another important difference between the ring of algebraic integers and the ring of rational integers. The concept of an irreducible integer (in analogy to a prime number) cannot be introduced into the former. This may be seen from the fact that a root of an algebraic integer is an algebraic integer. The concept of an irreducible number (apart from the class of associated numbers) can be introduced in certain subfields of the field of all algebraic numbers, the so-called algebraic number fields. It turns out, however, that the decomposition of an algebraic integer into irreducible factors is not always unique.

Algebraic numbers cannot be very closely approximated by rational and algebraic numbers (Liouville's theorem). It is this fact which led in 1844 to a proof of the existence of transcendental numbers. The problem of approximation of algebraic numbers by rational numbers is one of the more difficult problems in number theory; attempts to solve it yielded very important results, including the Thue, Thue–Siegel and Thue–Siegel–Roth theorems, but its ultimate solution is still nowhere in sight. Another very difficult problem is the expansion of algebraic numbers into continued fractions. Real algebraic numbers of the second degree (quadratic irrationalities) can be represented as infinite periodic continued fractions. Nothing is known so far about the expansion of real algebraic numbers of degree at least three into ordinary continued fractions.

A complex number is called an algebraic number over a field $P \subset C$ if it is a root of a polynomial ??? with coefficients from P. The minimal polynomial, the degree over P and the conjugate numbers over P for algebraic numbers over P are defined in a similar manner. A root of a polynomial with coefficients that are algebraic numbers over P is an algebraic number over P.

An algebraic number of arbitrary degree n does not necessarily exist over any field P. For instance, only algebraic numbers of the first degree exist over the field of complex numbers: the numbers of the field themselves. A given algebraic number can be of different degrees with respect to different fields; thus, the number ii is an algebraic number of the second degree, but it is of the first degree with respect to the field of complex numbers. The set of all algebraic numbers over a field P forms a number field.

Algebraic numbers, and algebraic number fields, were first systematically studied by C.F. Gauss (Gaussian numbers of the form $a + bi$, where a and b are rational numbers). Gauss developed the arithmetic of Gaussian integers as a base for the theory of biquadratic residues. In their study of the theory of cubic residues, C.G.J. Jacobi and F.G. Eisenstein created the arithmetic of numbers of the form $a + b\rho$, where $\rho = (-1 + \sqrt{3})/2$ is a cubic root of unity, while a and b are rational numbers. His attempts to produce a proof of the Fermat theorem led E. Kummer to conduct a deep study of cyclotomic fields (cf. Cyclotomic field), to introduce the concept of an ideal, and to create the elements of the theory of algebraic numbers. The theory of algebraic numbers was further developed by P. Dirichlet, L. Kronecker, D. Hilbert and others. Russian mathematicians — E.I. Zolotarev (theory of ideals), G.F. Voronoi (cubic irrationalities, units of cubic fields), A.A. Markov (cubic fields), Yu.V. Sokhotskii (theory of ideals) and others — also made significant contributions.

The concept of an algebraic number and the related concept of an algebraic number field are very important ideas in number theory and algebra. Algebraic numbers, which are a generalization of rational numbers, form subfields of algebraic numbers in the fields of real and complex numbers with special algebraic properties. The development of the theory of algebraic numbers greatly influenced the creation and development of the general theory of rings and fields.

Algebraic numbers have found numerous applications in various branches of number theory, algebra and other branches of mathematics: the theory of forms, Diophantine equations, Diophantine approximations, transcendental numbers, geometry of numbers, algebraic geometry, Galois theory, etc.

An algebraic integer is an algebraic number that is a root of a polynomial with integer coefficients with leading coefficient 1 (a monic polynomial).

with coefficients in $\mathbb{Z}$ (the set of integers). The set of all algebraic integers, A, is closed under addition, subtraction and multiplication and therefore is a commutative subring of the complex numbers. The ring A is the integral closure of regular integers $\mathbb{Z}$ in complex numbers.

The ring of integers of a number field K, denoted by O_K, is the intersection of K and A: it can also be characterised as the maximal order of the field K. Each algebraic integer belongs to the ring of integers of some number field. A number α is an algebraic integer if and only if the ring $\mathbb{Z}[\alpha]$ is finitely generated as an Abelian group, which is to say, as a $\mathbb{Z}$-module.

The following are equivalent definitions of an algebraic integer. Let K be a number field (i.e., a finite extension of $\mathbb{Q}$, the set of rational numbers), in other words, $K = \mathbb{Q}(\theta)$ for some algebraic number $\theta \in \mathbb{C}$ by the primitive element theorem.

- $\alpha \in K$ is an algebraic integer if there exists a monic polynomial $f(x) \in \mathbb{Z}[x]$ such that $f(\alpha) = 0$.

- $\alpha \in K$ is an algebraic integer if the minimal monic polynomial of α over $\mathbb{Q}$ is in $\mathbb{Z}[x]$.

- $\alpha \in K$ is an algebraic integer if $\mathbb{Z}[\alpha]$ is a finitely generated $\mathbb{Z}$-module.

- $\alpha \in K$ is an algebraic integer if there exists a non-zero finitely generated $\mathbb{Z}$-submodule $M \subset \mathbb{C}$ such that $\alpha M \subseteq M$.

Algebraic integers are a special case of integral elements of a ring extension. In particular, an algebraic integer is an integral element of a finite extension $K/\mathbb{Q}$.

Examples:

- The only algebraic integers which are found in the set of rational numbers are the integers. In other words, the intersection of $\mathbb{Q}$ and A is exactly $\mathbb{Z}$. The rational number $\dfrac{a}{b}$ is not an algebraic integer unless b divides a. Note that the leading coefficient of the polynomial $bx - a$ is the integer b. As another special case, the square root $\sqrt{n}$ of a nonnegative integer n is an algebraic integer, but is irrational unless n is a perfect square.

- If d is a square-free integer then the extension $K = \mathbb{Q}\left(\sqrt{d}\right)$ is a quadratic field of rational numbers. The ring of algebraic integers O_K contains $\sqrt{d}$ since this is a root of the monic polynomial $x^2 - d$. Moreover, if $d \equiv 1 \bmod 4$, then the element $\dfrac{1}{2}\left(1 + \sqrt{d}\right)$ is also an algebraic integer. It satisfies the polynomial $x^2 - x + \dfrac{1}{4}(1 - d)$ where the constant term $\dfrac{1}{4}(1 - d)$ is an integer. The full ring of integers is generated by $\sqrt{d}$ or $\dfrac{1}{2}\left(1 + \sqrt{d}\right)$ respectively. See quadratic integers for more.

- The ring of integers of the field $F = Q[\alpha], \alpha = \sqrt[3]{m}$, has the following integral basis, writing $m = hk^2$ for two square-free coprime integers h and k:

$$\begin{cases} 1, \alpha, \dfrac{\alpha^2 \pm k^2 \alpha + k^2}{3k} & m \equiv \pm 1 \bmod 9 \\[3mm] 1, \alpha, \dfrac{\alpha^2}{k} & \text{otherwise} \end{cases}$$

- If ζ_n is a primitive nth root of unity, then the ring of integers of the cyclotomic field $\mathbb{Q}(\zeta_n)$ is precisely $\mathbb{Z}[\zeta_n]$.

- If α is an algebraic integer then $\beta = \sqrt[n]{a}$ is another algebraic integer. A polynomial for β is obtained by substituting x^n in the polynomial for α.

Non-example:

- If $P(x)$ is a primitive polynomial which has integer coefficients but is not monic, and P is irreducible over $\mathbb{Q}$, then none of the roots of P are algebraic integers (but *are* algebraic numbers). Here *primitive* is used in the sense that the highest common factor of the set of coefficients of P is 1; this is weaker than requiring the coefficients to be pairwise relatively prime.

- The sum, difference and product of two algebraic integers is an algebraic integer. In general their quotient is not. The monic polynomial involved is generally of higher degree than those of the original algebraic integers, and can be found by taking resultants and factoring. For example, if $x^2 - x - 1$, $y^3 - y - 1$ and $z = xy$, then eliminating x and y from $z - xy$ and the polynomials satisfied by x and y using the resultant gives $z^6 - 3z^4 - 4z^3 + z^2 + z - 1$, which is irreducible, and is the monic polynomial satisfied by the product. (To see that the xy is a root of the x-resultant of $z - xy$ and $x^2 - x - 1$, one might use the fact that the resultant is contained in the ideal generated by its two input polynomials).

- Any number constructible out of the integers with roots, addition, and multiplication is therefore an algebraic integer; but not all algebraic integers are so constructible: in a naïve sense, most roots of irreducible quintics are not. This is the Abel–Ruffini theorem.

- Every root of a monic polynomial whose coefficients are algebraic integers is itself an algebraic integer. In other words, the algebraic integers form a ring which is integrally closed in any of its extensions.

- The ring of algebraic integers is a Bézout domain, as a consequence of the principal ideal theorem.

- If the monic polynomial associated with an algebraic integer has constant term 1 or -1, then the reciprocal of that algebraic integer is also an algebraic integer, and is a unit, an element of the group of units of the ring of algebraic integers.

ALGEBRAIC NUMBER THEORY

Algebraic number theory is a branch of number theory that uses the techniques of abstract algebra to study the integers, rational numbers, and their generalizations. Number-theoretic questions are expressed in terms of properties of algebraic objects such as algebraic number fields and their rings of integers, finite fields, and function fields. These properties, such as whether a ring admits unique factorization, the behavior of ideals, and the Galois groups of fields, can resolve questions of primary importance in number theory, like the existence of solutions to Diophantine equations.

Basic Notions

Failure of Unique Factorization

An important property of the ring of integers is that it satisfies the fundamental theorem of arithmetic, that every (positive) integer has a factorization into a product of prime numbers, and this factorization is unique up to the ordering of the factors. This may no longer be true in the ring of integers O of an algebraic number field K.

A *prime element* is an element p of O such that if p divides a product ab, then it divides one of the factors a or b. This property is closely related to primality in the integers, because any positive integer satisfying this property is either 1 or a prime number. However, it is strictly weaker. For example, -2 is not a prime number because it is negative, but it is a prime element. If factorizations into prime elements are permitted, then, even in the integers, there are alternative factorizations such as

$$6 = 2 \cdot 3 = (-2) \cdot (-3).$$

In general, if u is a unit, meaning a number with a multiplicative inverse in O, and if p is a prime element, then up is also a prime element. Numbers such as p and up are said to be *associate*. In the integers, the primes p and $-p$ are associate, but only one of these is positive. Requiring that prime numbers be positive selects a unique element from among a set of associated prime elements. When K is not the rational numbers, however, there is no analog of positivity. For example, in the Gaussian integers Z[i], the numbers $1 + 2i$ and $-2 + i$ are associate because the latter is the product of the former by i, but there is no way to single out one as being more canonical than the other. This leads to equations such as

$$5 = (1 + 2i)(1 - 2i) = (2 + i)(2 - i),$$

which prove that in Z[i], it is not true that factorizations are unique up to the order of the factors. For this reason, one adopts the definition of unique factorization used in unique factorization domains (UFDs). In a UFD, the prime elements occurring in a factorization are only expected to be unique up to units and their ordering.

However, even with this weaker definition, many rings of integers in algebraic number fields do not admit unique factorization. There is an algebraic obstruction called the ideal class group. When the ideal class group is trivial, the ring is a UFD. When it is not, there is a distinction between a prime element and an irreducible element. An *irreducible element* x is an element such that if $x = yz$, then either y or z is a unit. These are the elements that cannot be factored any further. Every element in O admits a factorization into irreducible elements, but it may admit more than one. This is because, while all prime elements are irreducible, some irreducible elements may not be prime. For example, consider the ring $Z[\sqrt{-5}]$. In this ring, the numbers 3, $2 + \sqrt{-5}$ and $2 - \sqrt{-5}$ are irreducible. This means that the number 9 has two factorizations into irreducible elements,

$$9 = 3^2 = (2 + \sqrt{-5})(2 - \sqrt{-5}).$$

This equation shows that 3 divides the product $(2 + \sqrt{-5})(2 - \sqrt{-5}) = 9$. If 3 were a prime element, then it would divide $2 + \sqrt{-5}$ or $2 - \sqrt{-5}$, but it does not, because all elements divisible by 3 are of the form $3a + 3b\sqrt{-5}$. Similarly, $2 + \sqrt{-5}$ and $2 - \sqrt{-5}$ divide the product 3^2, but neither of these elements divides 3 itself, so neither of them are prime. As there is no sense in which the elements 3, $2 + \sqrt{-5}$ and $2 - \sqrt{-5}$ can be made equivalent, unique factorization fails in $Z[\sqrt{-5}]$. Unlike the situation with units, where uniqueness could be repaired by weakening the definition, overcoming this failure requires a new perspective.

Factorization into Prime Ideals

If I is an ideal in O, then there is always a factorization

$$I = \mathfrak{p}_1^{e_1} \cdots \mathfrak{p}_t^{e_t},$$

where each $\mathfrak{p}_i$ is a prime ideal, and where this expression is unique up to the order of the factors. In particular, this is true if I is the principal ideal generated by a single element. This is the strongest sense in which the ring of integers of a general number field admits unique factorization. In the language of ring theory, it says that rings of integers are Dedekind domains.

When O is a UFD, every prime ideal is generated by a prime element. Otherwise, there are prime ideals which are not generated by prime elements. In $Z[\sqrt{-5}]$, for instance, the ideal $(2, 1 + \sqrt{-5})$ is a prime ideal which cannot be generated by a single element.

Historically, the idea of factoring ideals into prime ideals was preceded by Ernst Kummer's introduction of ideal numbers. These are numbers lying in an extension field E of K. This extension field is now known as the Hilbert class field. By the principal ideal theorem, every prime ideal of O generates a principal ideal of the ring of integers of E. A generator of this principal ideal is called an ideal number. Kummer used these as a substitute for the failure of unique factorization in cyclotomic fields. These eventually led Richard Dedekind to introduce a forerunner of ideals and to prove unique factorization of ideals.

An ideal which is prime in the ring of integers in one number field may fail to be prime when extended to a larger number field. Consider, for example, the prime numbers. The corresponding ideals pZ are prime ideals of the ring Z. However, when this ideal is extended to the Gaussian integers to get pZ[i], it may or may not be prime. For example, the factorization $2 = (1 + i)(1 - i)$ implies that

$$2Z[i] = (1+i)Z[i] \cdot (1-i)Z[i] = ((1+i)Z[i])^2;$$

note that because $1 + i = (1 - i) \cdot i$, the ideals generated by $1 + i$ and $1 - i$ are the same. A complete answer to the question of which ideals remain prime in the Gaussian integers is provided by Fermat's theorem on sums of two squares. It implies that for an odd prime number p, pZ[i] is a prime ideal if $p \equiv 3 \pmod 4$ and is not a prime ideal if $p \equiv 1 \pmod 4$. This, together with the observation that the ideal $(1 + i)Z[i]$ is prime, provides a complete description of the prime ideals in the Gaussian integers. Generalizing this simple result to more general rings of integers is a basic problem in algebraic number theory. Class field theory accomplishes this goal when K is an abelian extension of Q (i.e. a Galois extension with abelian Galois group).

Ideal Class Group

Unique factorization fails if and only if there are prime ideals that fail to be principal. The object which measures the failure of prime ideals to be principal is called the ideal class group. Defining the ideal class group requires enlarging the set of ideals in a ring of algebraic integers so that they admit a group structure. This is done by generalizing ideals to fractional ideals. A fractional ideal is an additive subgroup J of K which is closed under multiplication by elements of O, meaning that $xJ \subseteq J$ if $x \in O$. All ideals of O are also fractional ideals. If I and J are fractional ideals, then the set

IJ of all products of an element in *I* and an element in *J* is also a fractional ideal. This operation makes the set of non-zero fractional ideals into a group. The group identity is the ideal (1) = *O*, and the inverse of *J* is a (generalized) ideal quotient, $J^{-1} = (O : J) = \{ x \in K : xJ \subseteq O \}$.

The principal fractional ideals, meaning the ones of the form *Ox* where $x \in K^{\times}$, form a subgroup of the group of all non-zero fractional ideals. The quotient of the group of non-zero fractional ideals by this subgroup is the ideal class group. Two fractional ideals *I* and *J* represent the same element of the ideal class group if and only if there exists an element $x \in K$ such that $xI = J$. Therefore, the ideal class group makes two fractional ideals equivalent if one is as close to being principal as the other is. The ideal class group is generally denoted Cl *K*, Cl *O*, or Pic *O* (with the last notation identifying it with the Picard group in algebraic geometry).

The number of elements in the class group is called the class number of *K*. The class number of $\mathbf{Q}(\sqrt{-5})$ is 2. This means that there are only two ideal classes, the class of principal fractional ideals, and the class of a non-principal fractional ideal such as $(2, 1 + \sqrt{-5})$.

The ideal class group has another description in terms of divisors. These are formal objects which represent possible factorizations of numbers. The divisor group Div *K* is defined to be the free abelian group generated by the prime ideals of *O*. There is a group homomorphism from $K^{\times}$, the non-zero elements of *K* up to multiplication, to Div *K*. Suppose that $x \in K$ satisfies

$$(x) = \mathfrak{p}_1^{e_1} \cdots \mathfrak{p}_t^{e_t}.$$

Then div *x* is defined to be the divisor

$$\operatorname{div} x = \sum_{i=1}^{t} e_i [\mathfrak{p}_i].$$

The kernel of div is the group of units in *O*, while the cokernel is the ideal class group. In the language of homological algebra, this says that there is an exact sequence of abelian groups (written multiplicatively),

$$1 \to O^{\times} \to K^{\times} \xrightarrow{\operatorname{div}} \operatorname{Div} K \to \operatorname{Cl} K \to 1.$$

Real and Complex Embeddings

Some number fields, such as $\mathbf{Q}(\sqrt{2})$, can be specified as subfields of the real numbers. Others, such as $\mathbf{Q}(\sqrt{-1})$, cannot. Abstractly, such a specification corresponds to a field homomorphism $K \to \mathbf{R}$ or $K \to \mathbf{C}$. These are called real embeddings and complex embeddings, respectively.

A real quadratic field $\mathbf{Q}(\sqrt{a})$, with $a \in \mathbf{R}$, $a > 0$, and *a* not a perfect square, is so-called because it admits two real embeddings but no complex embeddings. These are the field homomorphisms which send $\sqrt{a}$ to $\sqrt{a}$ and to $-\sqrt{a}$, respectively. Dually, an imaginary quadratic field $\mathbf{Q}(\sqrt{-a})$ admits no real embeddings but admits a conjugate pair of complex embeddings. One of these embeddings sends $\sqrt{-a}$ to $\sqrt{-a}$, while the other sends it to its complex conjugate, $-\sqrt{-a}$.

Conventionally, the number of real embeddings of *K* is denoted r_1, while the number of conjugate pairs of complex embeddings is denoted r_2. The signature of *K* is the pair (r_1, r_2). It is a theorem that $r_1 + 2r_2 = d$, where *d* is the degree of *K*.

Considering all embeddings at once determines a function

$$M : K \to \mathbf{R}^{r_1} \oplus \mathbf{C}^{2r_2}.$$

This is called the Minkowski embedding. The subspace of the codomain fixed by complex conjugation is a real vector space of dimension d called Minkowski space. Because the Minkowski embedding is defined by field homomorphisms, multiplication of elements of K by an element $x \in K$ corresponds to multiplication by a diagonal matrix in the Minkowski embedding. The dot product on Minkowski space corresponds to the trace form $\langle x, y \rangle = \mathrm{Tr}(xy)$.

The image of O under the Minkowski embedding is a d-dimensional lattice. If B is a basis for this lattice, then $\det B^{\mathrm{T}}B$ is the discriminant of O. The discriminant is denoted Δ or D. The covolume of the image of O is $\sqrt{|\Delta|}$.

Places

Real and complex embeddings can be put on the same footing as prime ideals by adopting a perspective based on valuations. Consider, for example, the integers. In addition to the usual absolute value function $|\cdot| : \mathbf{Q} \to \mathbf{R}$, there are p-adic absolute value functions $|\cdot|_p : \mathbf{Q} \to \mathbf{R}$, defined for each prime number p, which measure divisibility by p. Ostrowski's theorem states that these are all possible absolute value functions on Q (up to equivalence). Therefore, absolute values are a common language to describe both the real embedding of Q and the prime numbers.

A place of an algebraic number field is an equivalence class of absolute value functions on K. There are two types of places. There is a p-adic absolute value for each prime ideal p of O, and, like the p-adic absolute values, it measures divisibility. These are called finite places. The other type of place is specified using a real or complex embedding of K and the standard absolute value function on R or C. These are infinite places. Because absolute values are unable to distinguish between a complex embedding and its conjugate, a complex embedding and its conjugate determine the same place. Therefore, there are r_1 real places and r_2 complex places. Because places encompass the primes, places are sometimes referred to as primes. When this is done, finite places are called finite primes and infinite places are called infinite primes. If v is a valuation corresponding to an absolute value, then one frequently writes $v \mid \infty$ to mean that v is an infinite place and $v \nmid \infty$ to mean that it is a finite place.

Considering all the places of the field together produces the adele ring of the number field. The adele ring allows one to simultaneously track all the data available using absolute values. This produces significant advantages in situations where the behavior at one place can affect the behavior at other places, as in the Artin reciprocity law.

Units

The integers have only two units, 1 and −1. Other rings of integers may admit more units. The Gaussian integers have four units, the previous two as well as $\pm i$. The Eisenstein integers $\mathbf{Z}[\exp(2\pi i / 3)]$ have six units. The integers in real quadratic number fields have infinitely many units. For example, in $\mathbf{Z}[\sqrt{3}]$, every power of $2 + \sqrt{3}$ is a unit, and all these powers are distinct.

In general, the group of units of O, denoted $O^\times$, is a finitely generated abelian group. The fundamental theorem of finitely generated abelian groups therefore implies that it is a direct sum of a torsion part and a free part. Reinterpreting this in the context of a number field, the torsion part consists of the roots of unity that lie in O. This group is cyclic. The free part is described by Dirichlet's unit theorem. This theorem says that rank of the free part is $r_1 + r_2 - 1$. Thus, for example, the only fields for which the rank of the free part is zero are Q and the imaginary quadratic fields. A more precise statement giving the structure of $O^\times \otimes_Z Q$ as a Galois module for the Galois group of K/Q is also possible.

The free part of the unit group can be studied using the infinite places of K. Consider the function

$$L : K^\times \to R^{r_1 + r_2}$$

defined by

$$L(x) = (\log |x|_v)_v,$$

where v varies over the infinite places of K and $|\cdot|_v$ is the absolute value associated with v. The function L is a homomorphism from $K^\times$ to a real vector space. It can be shown that the image of $O^\times$ is a lattice that spans the hyperplane defined by $x_1 + \cdots + x_{r_1 + r_2} = 0$. The covolume of this lattice is the regulator of the number field. One of the simplifications made possible by working with the adele ring is that there is a single object, the idele class group, that describes both the quotient by this lattice and the ideal class group.

Zeta Function

The Dedekind zeta function of a number field, analogous to the Riemann zeta function is an analytic object which describes the behavior of prime ideals in K. When K is an abelian extension of Q, Dedekind zeta functions are products of Dirichlet L-functions, with there being one factor for each Dirichlet character. The trivial character corresponds to the Riemann zeta function. When K is a Galois extension, the Dedekind zeta function is the Artin L-function of the regular representation of the Galois group of K, and it has a factorization in terms of irreducible Artin representations of the Galois group.

The zeta function is related to the other invariants described above by the class number formula.

Local Fields

Completing a number field K at a place w gives a complete field. If the valuation is archimedean, one gets R or C, if it is non-archimedean and lies over a prime p of the rationals, one gets a finite extension K_w/Q_p: a complete, discrete valued field with finite residue field. This process simplifies the arithmetic of the field and allows the local study of problems. For example, the Kronecker–Weber theorem can be deduced easily from the analogous local statement. The philosophy behind the study of local fields is largely motivated by geometric methods. In algebraic geometry, it is common to study varieties locally at a point by localizing to a maximal ideal. Global information can then be recovered by gluing together local data. This spirit is adopted in algebraic number theory. Given a prime in the ring of algebraic integers in a number field, it is desirable to study the field locally at that prime. Therefore, one localizes the ring of algebraic integers to that prime and then completes the fraction field much in the spirit of geometry.

Major Results

Finiteness of the Class Group

One of the classical results in algebraic number theory is that the ideal class group of an algebraic number field K is finite. The order of the class group is called the class number, and is often denoted by the letter h.

Dirichlet's Unit Theorem

Dirichlet's unit theorem provides a description of the structure of the multiplicative group of units $O^\times$ of the ring of integers O. Specifically, it states that $O^\times$ is isomorphic to $G \times Z^r$, where G is the finite cyclic group consisting of all the roots of unity in O, and $r = r_1 + r_2 - 1$ (where r_1 (respectively, r_2) denotes the number of real embeddings (respectively, pairs of conjugate non-real embeddings) of K). In other words, $O^\times$ is a finitely generated abelian group of rank $r_1 + r_2 - 1$ whose torsion consists of the roots of unity in O.

Reciprocity Laws

In terms of the Legendre symbol, the law of quadratic reciprocity for positive odd primes states

$$\left(\frac{p}{q}\right)\left(\frac{q}{p}\right) = (-1)^{\frac{p-1}{2}\frac{q-1}{2}}.$$

A reciprocity law is a generalization of the law of quadratic reciprocity.

There are several different ways to express reciprocity laws. The early reciprocity laws found in the 19th century were usually expressed in terms of a power residue symbol (p/q) generalizing the quadratic reciprocity symbol, that describes when a prime number is an nth power residue modulo another prime, and gave a relation between (p/q) and (q/p). Hilbert reformulated the reciprocity laws as saying that a product over p of Hilbert symbols $(a,b/p)$, taking values in roots of unity, is equal to 1. Artin's reformulated reciprocity law states that the Artin symbol from ideals (or ideles) to elements of a Galois group is trivial on a certain subgroup. Several more recent generalizations express reciprocity laws using cohomology of groups or representations of adelic groups or algebraic K-groups, and their relationship with the original quadratic reciprocity law can be hard to see.

Class Number Formula

The class number formula relates many important invariants of a number field to a special value of its Dedekind zeta function.

CLASS FIELD THEORY

In mathematics, class field theory is the branch of algebraic number theory concerned with the abelian extensions of number fields, global fields of positive characteristic, and local fields. The theory had its origins in the proof of quadratic reciprocity by Gauss at the end of 18th century.

These ideas were developed over the next century, giving rise to a set of conjectures by Hilbert that were subsequently proved by Takagi and Artin. These conjectures and their proofs constitute the main body of class field theory.

One major result states that, given a number field F, and writing K for the maximal abelian unramified extension of F, the Galois group of K over F is canonically isomorphic to the ideal class group of F. This statement can be generalized to the Artin reciprocity law; writing C_F for the idele class group of F, and taking L to be any finite abelian extension of F, this law gives a canonical isomorphism

$$\theta_{L/F} : C_F / N_{L/F}(C_L) \to \mathrm{Gal}(L / F),$$

where $N_{L/F}$ denotes the idelic norm map from L to F. This isomorphism is then called the *reciprocity map*. The *existence theorem* states that the reciprocity map can be used to give a bijection between the set of abelian extensions of F and the set of closed subgroups of finite index of C_F.

A standard method for developing global class field theory since the 1930s is to develop local class field theory, which describes abelian extensions of local fields, and then use it to construct global class field theory. This was first done by Artin and Tate using the theory of group cohomology, and in particular by developing the notion of class formations. More recently, Neukrich has found a proof of the main statements of global class field theory without using cohomological ideas.

Class field theory also encompasses the explicit construction of maximal abelian extensions of number fields in the few cases where such constructions are known. Currently, this portion of the theory consists of Kronecker-Weber theorem, which can be used to construct the abelian extensions of $\mathbb{Q}$ and the theory of complex multiplication, which can be used to construct the abelian extensions of CM-fields.

The Langlands program gives one approach for generalizing class field theory to non-abelian extensions. This generalization is mostly still conjectural. For number fields, class field theory and the results related to the modularity theorem are the only cases known.

Formulation in Contemporary Language

In modern mathematical language class field theory can be formulated as follows. Consider the *maximal* abelian extension A of a local or global field K. It is of infinite degree over K; the Galois group G of A over K is an infinite pro-finite group, so a compact topological group, and it is abelian. The central aims of class field theory are: to describe G in terms of certain appropriate topological objects associated to K, to describe finite abelian extensions of K in terms of open subgroups of finite index in the topological object associated to K. In particular, one wishes to establish a one-to-one correspondence between finite abelian extensions of K and their norm groups in this topological object for K. This topological object is the multiplicative group in the case of local fields with finite residue field and the idele class group in the case of global fields. The finite abelian extension corresponding to an open subgroup of finite index is called the class field for that subgroup, which gave the name to the theory.

The fundamental result of general class field theory states that the group G is naturally isomorphic to the profinite completion of C_K, the multiplicative group of a local field or the idele class group of the global field, with respect to the natural topology on C_K related to the specific structure of the field K. Equivalently, for any finite Galois extension L of K, there is an isomorphism (the Artin reciprocity map)

$$\mathrm{Gal}(L \,/\, K)^{\mathrm{ab}} \to C_K \,/\, N_{L/K}(C_L)$$

of the abelianization of the Galois group of the extension with the quotient of the idele class group of K by the image of the norm of the idele class group of L.

For some small fields, such as the field of rational numbers $\mathbb{Q}$ or its quadratic imaginary extensions there is a more detailed *very explicit but too specific* theory which provides more information. For example, the abelianized absolute Galois group G of $\mathbb{Q}$ is (naturally isomorphic to) an infinite product of the group of units of the p-adic integers taken over all prime numbers p, and the corresponding maximal abelian extension of the rationals is the field generated by all roots of unity. This is known as the Kronecker–Weber theorem, originally conjectured by Leopold Kronecker. In this case the reciprocity isomorphism of class field theory (or Artin reciprocity map) also admits an explicit description due to the Kronecker–Weber theorem. However, principal constructions of such more detailed theories for small algebraic number fields are not extendable to the general case of algebraic number fields, and different conceptual principles are in use in the general class field theory.

The standard method to construct the reciprocity homomorphism is to first construct the local reciprocity isomorphism from the multiplicative group of the completion of a global field to the Galois group of its maximal abelian extension (this is done inside local class field theory) and then prove that the product of all such local reciprocity maps when defined on the idele group of the global field is trivial on the image of the multiplicative group of the global field. The latter property is called the *global reciprocity law* and is a far reaching generalization of the Gauss quadratic reciprocity law.

One of the methods to construct the reciprocity homomorphism uses class formation which derives class field theory from axioms of class field theory. This derivation is purely topological group theoretical, while to establish the axioms one has to use the ring structure of the ground field.

There are methods which use cohomology groups, in particular the Brauer group, and there are methods which do not use cohomology groups and are very explicit and fruitful for applications.

Applications

Class field theory is used to prove Artin-Verdier duality. Very explicit class field theory is used in many subareas of algebraic number theory such as Iwasawa theory and Galois modules theory.

Most main achievements in the Langlands correspondence for number fields, the BSD conjecture for number fields, and Iwasawa theory for number fields are using very explicit but narrow class field theory methods or their generalizations. The open question is therefore to use generalizations of general class field theory in these three directions.

Generalizations of Class Field Theory

There are three main generalizations, each of great interest on its own. They are: the Langlands program, anabelian geometry, and higher class field theory.

Often, the Langlands correspondence is viewed as a nonabelian class field theory. If/when fully established, it would contain a certain theory of nonabelian Galois extensions of global fields. However, the Langlands correspondence does not include as much arithmetical information about finite Galois extensions as class field theory does in the abelian case. It also does not include an analog of the existence theorem in class field theory, i.e. the concept of class fields is absent in the Langlands correspondence. There are several other nonabelian theories, local and global, which provide alternative to the Langlands correspondence point of view.

Another generalization of class field theory is anabelian geometry which studies algorithms to restore the original object (e.g. a number field or a hyperbolic curve over it) from the knowledge of its full absolute Galois group of algebraic fundamental group.

ABSTRACT ANALYTIC NUMBER THEORY

The central concept in abstract analytic number theory is that of an arithmetical semi-group G (defined below). It turns out that the study of such semi-groups and of (real- or complex-valued) functions on them makes it possible on the one hand to apply methods of classical analytic number theory in a unified way to a variety of asymptotic enumeration questions for isomorphism classes of different kinds of explicit mathematical objects. On the other hand, these procedures also lead to abstract generalizations and analogues of ordinary analytic number theory, which may then be applied in a unified way to further enumeration questions about the (mostly non-arithmetical) concrete types of mathematical objects just alluded to.

Arithmetical Semi-groups

An arithmetical semi-group is, by definition, a commutative semi-group G with identity element 1, which contains a countable subset P such that every element a $\neq$1 in G admits a unique factorization into a finite product of powers of elements of PP, together with a real-valued mapping $|\cdot|$ on G such that:

   i. $|1|=1, |p|>1$ for $p \in P$;

   ii. $|ab|=|a|\cdot|b|$ for all $a,b \in G$;

   iii.   The total number of elements a with $|a|< x$ is finite, for each $x > 0$.

The elements of P are called the primes of G, and $|\cdot|$ is called the norm mapping on G. It is obvious that, corresponding to any fixed c>1, the definition $\partial(a) = \log_c |a|$ yields a mapping on G such that:

   i. $\partial(1) = 0, \partial(p) > 0$ for $p \in P$;

   ii. $\partial(ab) = \partial(a) + \partial(b)$ for all $a,b \in G$;

   iii.   The total number of elements a with $\partial(a) \le x$ is finite, for each $x > 0$.

Conversely, any real-valued mapping ∂ with the properties A)–C) yields a norm on G, if one defines $|a|= c^{\partial(a)}$. In cases where such a mapping ∂ is of primary interest, G together with ∂ is called an additive arithmetical semi-group, and one refers to ∂ as the degree mapping on G. In most concrete examples of interest, it turns out that the norm or degree mappings represent natural "size" or "dimension" measures which are integer-valued. With an eye to applications to natural examples there is therefore little loss in henceforth restricting attention to either a single integer-valued norm mapping $|\cdot|$, or a single integer-valued degree mapping ∂, on G. Depending on which case is being considered, special interest then attaches to the basic counting functions (for $n \in Z$)

$$G(n) = \#\{a \in G : |a| = n\}$$
$$P(n) = \#\{p \in P : |p| = n\}$$

or $G^{\#}(n) = \#\{a \in G : \partial(a) = n\}, P^{\#}(n) = \#\{p \in P : \partial(p) = n\}$ in the additive case).

The prototype of all arithmetical semi-groups is of course the multiplicative semi-group $\mathbb{N}$ of all positive integers $\{1,2,3,...\}$ with its subset P_N of all rational prime numbers $\{2,3,5,7,...\}$ Here one may define the norm of an integer $|n|$ to be n, so that the number $\mathbb{N}(n) = 1$ for $n \geq 1$.

The asymptotic behaviour of $\pi(X) = \sum_{n \leq x} P_N(n)$ for large xx forms the content of the famous prime number theorem, which states that

$$\pi(x) \sim \frac{x}{\log x} \text{ as } x \to \infty$$

A suitably generalized form of this theorem holds for many other naturally-occurring arithmetical semi-groups. For example, it is true for the multiplicative semi-group G_K of all non-zero ideals in the ring $R = R(K)$ of all algebraic integers in a given algebraic number field K, with $|I| = card(R / I)$ for any non-zero ideal I in R. Here, the prime ideals act as prime elements of the semi-group G_K.

A simple but nevertheless interesting example of an additive arithmetical semi-group is provided by the multiplicative semi-group $G-q$ of all monic polynomials in one indeterminate X over a finite field $\mathbb{F}_q$ with q elements, with $\partial a = \deg a$ and the set $P q$ of prime elements represented by the irreducible polynomials. Here, $G_q^{\#}(n) = q^n$, and it can be proved that

$$P_q^{\#}(n) = \frac{1}{n} \sum_{r|n} \mu(r) q^{n/r}$$

where μ is the classical Möbius function on $\mathbb{N}$.

Up to isomorphism, G_q is the simplest special case of the semi-group G_R of all non-zero ideals in the ring $R=R(K)$ of all integral functions in an algebraic function field K in one variable K over $\mathbb{F}_q$.

Arithmetical Categories of Semi-groups

Many interesting examples of concrete, but non-classical, arithmetical semi-groups can be found by considering certain specific classes of mathematical objects, such as groups, rings, topological spaces, and so on, together with appropriate "direct product" operations and isomorphism relations for those classes. It is convenient, though admittedly not quite precise, to temporarily ignore the corresponding morphisms and refer to such classes of objects as "categories".

Now consider some category C which admits a direct "product" (or "sum") operation × on its objects. Suppose that this operation × preserves C-isomorphisms, is commutative and associative up to C-isomorphism, and that C contains a "zero" object o (unique up to C-isomorphism) such that $A \times 0 \cong A$ for all objects A in C. Then suppose that a theorem of Krull–Schmidt type is valid for C, i.e., suppose that every object A can be expressed as a finite ×-product $A \cong P_1 \times \cdots \times P_k$ of objects P_i that are indecomposable with respect to ×, in a way that is unique up to permutation of terms and C-isomorphism. In most natural situations at least, one may reformulate these conditions on C by stating that the various isomorphism classes $\overline{A}$ of objects A in C form a set G_C that is

i.   a commutative semi-group with identity with respect to the multiplication operation $\overline{A} \times \overline{B} = \overline{A \times B}$;

ii. a semi-group with the unique factorization property with respect to the isomorphism classes of the indecomposable objects in C.

For this reason, one may call the C-isomorphism classes $\overline{P}_i$ of indecomposable objects P the "primes" of C or G_C.

In many interesting cases (some of which are illustrated below), the category C also admits a "norm" function $|\cdot|$ on objects which is invariant under C-isomorphism and has the following properties:

i. $|0| = 1$, $|P| > 1$ for every indecomposable object P;

ii. $|A \times B| = |A| . |B|$ for all objects A, B;

iii. the total number of C-isomorphism classes of objects A of norm $|A| \leq x$ is finite, for each real $x > 0$.

Obviously, in such circumstances, the definition $|\overline{A}| = |A|$ provides a norm function on G_C satisfying the required conditions for an arithmetical semi-group. For these reasons, a category C with such further properties may be called an arithmetical category.

Now consider some concrete illustrations for the above concepts, taken from [a2], [a3]:

- One of the simplest non-trivial examples of an arithmetical category is provided by the category A of all finite Abelian groups, together with the usual direct product operation and the norm function $|A|$=card(A). Here, the Krull–Schmidt theorem reduces to the well-known fundamental theorem on finite Abelian groups, the indecomposable objects of this kind being simply the various cyclic groups C_{p^n} of prime-power order p^n.

- The category of all semi-simple associative rings of finite cardinality.

- The category of all semi-simple finite-dimensional associate algebras over a given field FF.

- The category of all semi-simple finite-dimensional Lie algebras over a given field F.

- The category of all compact simply-connected globally symmetric Riemannian manifolds.

- The category T of topological spaces of finite cardinality with the property that a space Y lies in T if and only if each connected component of Y lies in T.

Zeta-functions and Enumeration Problems

For a given arithmetical semi-group G, information on the basic counting functions $G(n), P(n)$ can often be obtained, algebraically or with the aid of analysis, via a certain series-production relation called the Euler product formula for G.

Indeed, ignoring questions of convergence for the moment, note that (by the unique factorization into prime elements of G) the series

$$\zeta_G(z) = \sum_{n=1}^{\infty} G(n) n^{-z} = \sum_{\alpha \in G} |\alpha|^{-z} =$$

$$= 1 + \sum_{\substack{\text{all products } p_1^{r_1} \cdots p_m^{r_m} \\ \text{with } p_i \in p,\ r_i\, m \in \mathbb{N}}} \left| p_1^{r_1} \cdots p_m^{r_m} \right|^{-z} =$$

$$= 1 + \sum |p_1|^{-r_1 z} \cdots |p_m|^{-r_m z} =$$

$$= \prod_{p \in P} \left(1 - |p|^{-z} + |p|^{-2z} + \ldots \right) =$$

$$= \prod_{p \in P} \left(1 - |p|^{-z} \right)^{-1} =$$

$$= \prod_{m=2}^{\infty} \left(1 - m^{-z} \right)^{-P\langle m \rangle}.$$

As a function of z, $\zeta_G(z)$ is called the zeta-function of G. If G is an additive arithmetical semi-group with $|\alpha| = c^{\partial\langle \alpha \rangle}$ for some integer $c > 1$, one may substitute the symbol Y for c^{-z} and obtain the modified Euler product formula:

$$\sum_{n=0}^{\infty} G^{\#}(n) y^n = \prod_{m=1}^{\infty} \left(1 - y^m \right) - P^{\#\langle m \rangle};$$

Then $Z_G(y) = \sum_{n=0}^{\infty} G^{\#}(n) y^n$ is called the modified zeta-function (or generating function) of G. Some explicit illustrations of zeta-functions and Euler products are given below.

The Riemann Zeta-function

For the basic semi-group N of positive integers, the zeta-function is

$$\zeta(z) = \sum_{n=1}^{\infty} n^{-z};$$

it is called the Riemann zeta-function, and the classical Euler product formula reads:

$$\zeta(z) = \prod_{\text{primes } p \text{ of } N} \left(1 - p^{-z} \right)^{-1}$$

The Dedekind Zeta-function

Let G_k denote the (above-mentioned) arithmetical semi-group of all non-zero "integral" ideals in a given algebraic number field K. The zeta-function for G_k is then

where $K(n)$ denotes the total number of ideals of norm n in G_k; it is known as the Dedekind zeta-function of K.

$$\zeta_K(z) = \sum_{I \in G_K} |I|^{-z} = \sum_{n=1}^{\infty} K(n) n^{-z},$$

Monic Polynomials Over a Finite Field

For the additive arithmetical semi-group G_q of all monic polynomials in one indeterminate X over Fq, the generating function may be written as:

$$Z_q(y) = \sum_{n=0}^{\infty} q^n y^n = (1 - qy)^{-1},$$

and the above-mentioned explicit formula for $P_q^{\#}(n)$ can be deduced as an algebraic consequence of the Euler product for G_q.

Finite Abelian Groups

For the category A of all finite Abelian groups, the zeta-function may be written as:

$$\zeta_A(z) = \sum_{n=1}^{\infty} \alpha(n) n^{-z},$$

where $\alpha(n)$ denotes the total number of isomorphism classes of Abelian groups of order n. The discussion of "primes" in A given above shows that here the Euler product may be written as a double product,

$$\zeta_A(z) = \prod_{\substack{r \geq 1, \\ primes\, p \in N}} \left(1 - p^{-rz}\right)^{-1} = \prod_{r=1}^{\infty} \zeta(rz),$$

by the Euler product formula for the Riemann zeta-function.

For the subcategory $A(p)$ of all finite Abelian P-groups, where P is a fixed prime number (cf. also P-group), it is natural to regard $A(p)$ as an additive arithmetical category, with degree mapping defined by,

$$\partial(A) = \log_p \operatorname{card}(A).$$

In that case, $A(p)$ has exactly one prime of degree r for each $r = 1, 2, \dots$. Therefore the Euler product formula implies that $A(p)$ has the generating function,

$$Z_{A\langle p \rangle}(y) = \prod_{r=1}^{\infty} \left(1 - y^r\right)^{-1} = \sum_{n=o}^{\infty} \mathrm{p}(n) y^n,$$

where $p(n)=\alpha\left(p^{n}\right)$ is the total number of isomorphism classes of Abelian groups of degree n in the above sense. In fact, for $n>0. p(n)$ equals the total number of ways of partitioning n into a sum of positive integers, which is also the number of pseudo-metrizable finite topological spaces of cardinality n. Thus, the corresponding latter category P (say) has the same generating function as $\mathcal{A}(p)$.

Types of Arithmetical Semi-groups

Bearing in mind the emphasis on concrete realizations of arithmetical semi-groups in a variety of areas of mathematics, it is reasonable to classify them and to base further investigations according to common features which may be exhibited by the initial enumeration theorems for particular sets of examples. In that way, further questions and enumeration problems may be investigated uniformly under suitable covering assumptions or "axioms" appropriate for particular natural sets of examples. On this basis, a small number of special types of arithmetical semi-groups have so far (2000) been found to predominate amongst natural concrete examples.

Classical and Axiom-*A* Type Semi-groups

The strictly classical arithmetical semi-groups of analytic number theory are the multiplicative semi-group of all positive integers and the multiplicative semi-group of all non-zero ideals in the ring of all algebraic integers in a given algebraic number field. For example, H. Weber and E. Landau proved theorems to the effect that

$$\sum_{n\leq x} G_k(n)=A_{Kx}+O\left(x^{\eta K}\right)\text{as }x\to\infty,$$

where G_K is the semi-group of all "integral" ideals in a given algebraic number field K. Landau in particular used (a1) in order to extend many asymptotic results about arithmetical functions on N to similar functions on G_K.

In quite a different direction, P. Erdős and G. Szekeres proved in 1934 for the category $\mathcal{A}$ of all finite Abelian groups that

$$\sum_{n\leq x}\alpha(n)=A_1 x+O\left(\sqrt{x}\right)\text{as }x\to\infty,$$

where $A_1=\prod_{r\leq2}\zeta(r)=2.29....$

At a later stage, for the category S of semi-simple finite rings, I.G. Connell and J. Knopfmacher independently proved that

$$\sum_{n\leq x}S(n)=A_2x+O\left(\sqrt{x}\right)\text{as }x\to\infty,$$

where $A_2=\prod_{rm^2\geq2}\zeta\left(rm^2\right)=2.49......$

Strong concrete motivation was available for unifying certain further developments under the umbrella of general studies of an abstract arithmetical semi-group G satisfying the so-called axiom A: There exist constants $A_G>0, \delta>0$ and $\eta<\delta$ (all depending on G), such that

$$\sum_{n \le x} G(n) = A_{Gx}{}^{\delta} + O\left(x^{\eta}\right) \text{as } x \to \infty.$$

Theorems based on the assumption of axiom A often simultaneously generalize earlier results for N.G_K and G_A, and provide additional asymptotic enumeration theorems for a variety of arithmetical categories like S and many others.

Axiom $A^{\#}$ Type Semi-groups

Consideration of the examples of multiplicative semi-groups of monic polynomials in one indeterminate, and also of enumeration theorems for some infinite families of explicit additive arithmetical categories connected with rings of integral functions in algebraic function fields over F_q, provides a wealth of motivation for studying an abstract additive arithmetical semi-group G satisfying axiom $A^{\#}$: There exist constants $A_G > 0. q > 1$ and $V < 1$ (all depending on G) such that

$$G^{\#}(n) = A_{G_q}{}^{n} + O\left(q^{vn}\right) \text{as } n \to \infty.$$

With this axiom as a basis instead of axiom A, problems similar to those outlined above may be investigated, with similar motivation to those stimulating the axiom- A type studies. It then turns out that the ensuing results and methods of proof sometimes but not always possess parallels to those subject to axiom A.

A curious illustration of a non-parallel result arises with the abstract prime number theorem (or abstract prime element theorem) subject to axiom $A^{\#}$. In 1976, Knopfmacher derived such a theorem, on the initial foundation of some plausible-looking lemmas parallel to ones under axiom A. However, in 1989 and later, other authors independently found and then closed certain gaps in those lemmas. The combined efforts of various authors then led to a final theorem with two cases, depending on whether or not $Z_G\left(-q^{-1}\right) = 0$; contributions to this were made by S.D. Cohen, K.-H. Indlekofer, E. Manstavičius, R. Warlimont and W.-B. Zhang.

A strange point about this result is that the case $Z_G\left(-q^{-1}\right) \neq 0$ holds for all the natural examples which initially motivated axiom $A^{\#}$. Although ingenious examples in which $Z_G\left(-q^{-1}\right) = 0$ have also been constructed, those found up to now might be viewed as somewhat pathological or contrived. Therefore, in terms of the "natural-example-based approach" to this subject outlined in the beginning, it would not be unreasonable to continue the present (2000) direction of investigation under the combined assumption of axiom $A^{\#}$ with the additional axiom

$$Z_G\left(-q^{-1}\right) \neq 0.$$

In fact many consequences of axiom $A^{\#}$ are unrelated to the value of $Z_G\left(-q^{-1}\right)$, and so the simplifying additional axiom would only sometimes become relevant (but nevertheless reasonable to then assume at such a stage).

Axiom C

The examples listed earlier included many involving an additive arithmetical category C for which $G_C^{\#}(n)$ and $P_C^{\#}(n)$ have quite a different behaviour from that given by axiom $A^{\#}$. Here, although the objects in C may sometimes be rather complicated, the presently (as of 2000) known structure theorems for those objects often lead to a relatively simple estimation for $P_C^{\#}(n)$ or $\pi_C^{\#}(x) = \sum_n \leq x P_C^{\#}(n)$. Surprisingly perhaps, it turns out that sharp asymptotic information can then be deduced about $G_C^{\#}(n)$ or $N_C^{\#}(x) = \sum_{n \leq x} G_C^{\#}(n)$ by methods of classical-type arithmetical partition theory, which were initiated by G.H. Hardy and G. Ramanujan in 1917. These methods belong to a quite different branch of classical analytic number theory from those involved in the earlier discussion of axiom A.

On the basis of these new types of examples as motivation, one is led to investigations of an additive arithmetical semi-group G satisfying axiom C: There exist constants, $C > 0. \kappa > 0$ and v (all depending on G) such that

$$\pi_C^{\#}(x) \sim Cx^{\kappa} (\log x)^v \text{ as } x \to \infty.$$

A simple example of axiom C is provided when C denotes either the category $\mathcal{A}(p)$ 3of finite Abelian P-groups, or the category $\mathcal{P}$ of pseudo-metrizable finite topological spaces.

Similar formulas hold for the categories of compact simply-connected Lie groups, or semi-simple finite-dimensional Lie algebras over an algebraically closed field F of characteristic zero.

Asymptotic deductions about $G^{\#}(n)$ or $N_G^{\#}(x) = \sum_n \leq x G^{\#}(n)$, subject to axiom C, could perhaps be referred to as "inverse additive abstract prime number theorems" . Based on methods of generalized arithmetical partition theory, various theorems of this kind can be derived, as well as results about "average values" of arithmetical functions on G, and on asymptotic "densities" of certain subsets of G, subject to axiom C.

Axiom G_1

Yet another natural class of additive arithmetical semi-groups G is provided by those satisfying axiom G_1: "Almost all" elements of G are prime, in the sense that $G^{\#}(n) > 0$ for sufficiently large n, and $P^{\#}(n) \sim G^{\#}(n)$ as $n \to \infty$.

It is known that various classes Γ of finite graphs define arithmetical semi-groups with this slightly surprising property. It is also known that, when $k > 1$, the multiplicative semi-group $G_{k,q}$ of all monic polynomials in k indeterminates $X_1, .., X_K$ over a finite field F_q has the property stipulated in axiom

Algebraic in Theorems

There are a number of important theorems used in algebra such as Abel–Ruffini theorem, Amitsur–Levitzki theorem, Bernstein–Kushnirenko theorem, Hilbert's basis theorem, remainder theorem and factor theorem. The topics elaborated in this chapter will help in gaining a better perspective about these theorems in algebra.

FUNDAMENTAL THEOREM OF ALGEBRA

The Fundamental theorem of algebra states that any nonconstant polynomial with complex coefficients has at least one complex root. The theorem implies that any polynomial with complex coefficients of degree n has n complex roots, counted with multiplicity. A field F with the property that every nonconstant polynomial with coefficients in F has a root in F is called algebraically closed, so the fundamental theorem of algebra states that

The field C of complex numbers is algebraically closed.

The polynomial $x^2 + 1$ has no real roots, but it has two complex roots i and -.-i.

The polynomial $x^2 + i$ has two complex roots, namely $\pm \dfrac{1-i}{\sqrt{2}}$.

One might expect that polynomials with complex coefficients have issues with nonexistence of roots similar to those of real polynomials; that is, it is not unreasonable to guess that some polynomial like

$$x^3 + ix^2 - (1 + \pi i)x - e$$

will not have a complex root, and finding such a root will require looking in some larger field containing the complex numbers. The fundamental theorem of algebra says that this is not the case: all the roots of a polynomial with complex coefficients can be found living inside the complex numbers already.

Factoring

This section gives a more precise statement of the different equivalent forms of the fundamental theorem of algebra. This requires a definition of the multiplicity of a root of a polynomial.

The multiplicity of a root r of a polynomial $f(x)$ is the largest positive integer k such that $(x-r)^k$ divides $f(x)$. Equivalently, it is the *smallest* positive integer k such that $f^{(k)}(r) \neq 0$, where $f^{(k)}$ denotes the kth derivative of f.

Theorem

Let F be a field. The following are equivalent:

1. Every nonconstant polynomial with coefficients in F has a root in F.

2. Every nonconstant polynomial of degree n with coefficients in F has n roots in F, counted with multiplicity.

3. Every nonconstant polynomial with coefficients in F splits completely as a product of linear factors with coefficients in F.

Proof:

Clearly $(3) \Rightarrow (2) \Rightarrow (1)$, so the only nontrivial part is $(1) \Rightarrow (3)$. To see this, induct on the degree n of $f(x)$. The base case $n=1$ is clear. Now suppose the result holds for polynomials of degree $n-1$. Then let $f(x)$ be a polynomial of degree n. By (1), $f(x)$ has a root $a.a$. A standard division algorithm argument shows that $x-a$ is a factor of $f(x)$.

Divide $f(x)$ by $x-a$ to get $f(x) = (x-a)q(x)+r$, where r is a constant polynomial. Plugging in a to both sides gives $0 = (a-a)q(a)+r$, so $r = 0$. So $f(x) = (x-a)q(x)$ But $q(x)$ is a polynomial of degree $n-1$, so it splits into a product of linear factors by the inductive hypothesis. Therefore $f(x)$ does as well. So the result is proved by induction.

The fundamental theorem of algebra says that the field \mathbb CC of complex numbers has property (1), so by the theorem above it must have properties (1), (2), and (3).

Example:

If $f(x) = x^4 - x^3 - x + 1$, then complex roots can be factored out one by one until the polynomial is factored completely: $f(1) = 0$, so $x^4 - x^3 - x + 1 = (x-1)(x^3 - 1)$. Then 1 is a root of $x^3 - 1$, so

$$x^4 - x^3 - x + 1 = (x-1)(x-1)(x^2 + x + 1),$$

and now $x^2 + x + 1$ has two complex roots, namely the primitive third roots of unity ω and ω^2, where $\omega = e^{2\pi i/3}$. So

$$x^4 - x^3 - x + 1 = (x-1)^2 (x-\omega)(x-\omega^2).$$

There are three distinct roots, but four roots with multiplicity, since the root 1 has multiplicity 2.

Applications of the Theorem

The ability to factor any polynomial over the complex numbers reduces many difficult nonlinear problems over other fields (e.g. the real numbers) to linear ones over the complex numbers. For example, every square matrix over the complex numbers has a complex eigenvalue, because the characteristic polynomial always has a root. This is not true over the real numbers, e.g. the matrix

$$\begin{pmatrix} 0 & 1 \\ -1 & 0 \end{pmatrix},$$

which rotates the real coordinate plane by $90°$, has no real eigenvalues.

Another general application is to the field of algebraic geometry, or the study of solutions to polynomial equations. The assumption that the coefficients of the polynomial equations lie in an algebraically closed field is essential for simplifying and strengthening the theory, as it guarantees that the field is "big enough" to contain roots of polynomials. For example, the set of complex solutions to a polynomial equation with real coefficients often has more natural and useful properties than the set of real solutions.

Another application worth mentioning briefly is to integration with partial fractions. Over the real numbers, there are awkward cases involving irreducible quadratic factors of the denominator. The algebra is simplified by using partial fractions over the complex numbers (with the caveat that some complex analysis is required to interpret the resulting integrals).

Polynomials over the Real Numbers

Let $p(x)$ be a polynomial with real coefficients. It is true that $p(x)$ can be factored into linear factors over the complex numbers, but the factorization is slightly more complicated if the factors are required to have real coefficients.

For instance, the polynomial $x^2 + 1$ can be factored as $(x-i)(x+i)$ over the complex numbers, but over the real numbers it is irreducible: it cannot be written as a product of two nonconstant polynomials with real coefficients.

Theorem

Every polynomial $p(x)$ with real coefficients can be factored into a product of linear and irreducible quadratic factors with real coefficients.

Proof:

Induct on n. The base cases are $n=1$ and $,n=2$, which are trivial. Now suppose the theorem is true for polynomials of degree $n-2$ and $n-1$. Let $f(x)$ be a polynomial of degree n, and let a be a complex root of $f(x)$ (which exists by the fundamental theorem of algebra). There are two cases.

If a is real, then $f(x) = (x-a)\,q(x)$ for a polynomial $q(x)$ with real coefficients of degree $n-1$. By the inductive hypothesis, $q(x)$ can be factored into a product of linear and irreducible quadratic factors, so $f(x)$ can as well.

If a is not real, then let $\bar{a}$ be the complex conjugate of a. Note that $\bar{a} \neq a$. Write $f(x) = f(x) = c_n x^n + \cdots + c_1 x + c_0$, then

$$\overline{f(x)} = \overline{c_n x^n + \cdots + c_1 x + c_0}$$
$$= \overline{c_n x^n} + \cdots + \overline{c_1 x} + \overline{c_0}$$
$$= \overline{c_n}\, \overline{x}^n + \cdots + \overline{c_1}\, \overline{x} + \overline{c_0}$$
$$= c_n \overline{x}^n + \cdots + c_1 \overline{x} + c_0$$
$$= f(\overline{x})$$

by properties of the complex conjugate (and because the c_i are real numbers). So if $f(a) = 0$, then o $f(\bar{a}) = \overline{f(a)} = \bar{0} = 0$. The conclusion is that non-real roots of polynomials with real coefficients come in complex conjugate pairs.

Write $f(x) = (x - a)q(x)$, where $q(x)$ has complex coefficients, and plug in $\bar{a}$ to both sides. Then $q(\bar{a}) = 0$. (This is where the argument uses that $\bar{a} \neq a$.) So $q(x) = (x - \bar{a})h(x)$, so $q(x) = (x - \bar{a})h(x)$, so $f(x) = (x - a)(x - \bar{a})h(x)$. Write the product of the first two factors as $g(x)$, then $g(x)$ is a quadratic irreducible polynomial with real coefficients. Since $g(x)$ divides $f(x)$ over the complex numbers, and both polynomials are real, $g(x)$ must divide $f(x)$ over the real numbers. (Proof: use the division algorithm over the real numbers, $f = f = g\,j + k$, with k =o or $(k) < \deg(g)$, and then over the complex numbers g divides f and $,gj$, so must divide k; so k=o).

So $h(x)$ is a polynomial of degree $n-2$ with real coefficients, which factors as expected by the inductive hypothesis, so $f(x)$ does as well. This completes the proof.

Proof of the Theorem

There are no "elementary" proofs of the theorem. The easiest proofs use basic facts from complex analysis. Here is a proof using Liouville's theorem that a bounded holomorphic function on the entire plane must be constant, along with a basic fact from topology about compact sets.

Let $p(z) = a_n z^n + \cdots + a_0$ be a polynomial with complex coefficients, and suppose that $p(z) \neq 0$ everywhere. (So of course a $a_0 \neq 0$.) Then $\dfrac{1}{p(z)}$ is holomorphic everywhere.

Now $\lim\limits_{z \to \infty} p(z) = \infty$. for instance, because

$$|p(z)| \geq |a_n||z|^n - \left(|a_{n-1}||z|^{n-1} + \cdots + |a_0|\right)$$

by the triangle inequality. So for large enough $|z|$, say $|z| > R$, $|p(z)| > |a_0|$.

But in the disc $|z| \leq R$, the function $|p(z)|$ attains its minimum value (because the disc is compact). Call this value m. Note that m>o.

Then $|p(z)| > \min\left(m, |a_0|\right)$ for all z, so

$$\left|\frac{1}{p(z)}\right| < \frac{1}{\min\left(m, |a_0|\right)}$$

for all z, so it is a bounded holomorphic function on the entire plane, so it must be constant by Liouville's theorem. But then $p(z)$ is constant.

So the argument has shown that any nonconstant polynomial with complex coefficients has a complex root, as desired.

POLYNOMIAL REMAINDER THEOREM

In algebra, the polynomial remainder theorem or little Bézout's theorem (named after Étienne Bézout) is an application of Euclidean division of polynomials. It states that the remainder of the division of a polynomial $f(x)$ by a linear polynomial $x - r$ is equal to $f(r)$ In particular, $x - r$ is a divisor of $f(x)$ if and only if $f(r) = 0$ a property known as the factor theorem.

Examples:

1. Let $f(x) = x^3 - 12x^2 - 42$. Polynomial division of $f(x)$ by $(x - 3)$ gives the quotient $x^2 - 9x - 27$ and the remainder -123. Therefore, $f(3) = -123$.

2. Show that the polynomial remainder theorem holds for an arbitrary second degree polynomial $f(x) = ax^2 + bx + c$ by using algebraic manipulation:

$$\frac{f(x)}{x-r} = \frac{ax^2 + bx + c}{x-r}$$

$$= \frac{ax^2 - arx + arx + bx + c}{x-r}$$

$$= \frac{ax(x-r) + (b+ar)x + c}{x-r}$$

$$= ax + \frac{(b+ar)(x-r) + c + r(b+ar)}{x-r}$$

$$= ax + b + ar + \frac{c + r(b+ar)}{x-r}$$

$$= ax + b + ar + \frac{ar^2 + br + c}{x-r}$$

Multiplying both sides by $(x - r)$ gives

$$f(x) = ax^2 + bx + c = (ax + b + ar)(x-r) + ar^2 + br + c.$$

Since $R = ar^2 + br + c$ is the remainder, we have indeed shown that $f(r) = R$.

Proof:

The polynomial remainder theorem follows from the theorem of Euclidean division, which, given two polynomials $f(x)$ (the dividend) and $g(x)$ (the divisor), asserts the existence (and the uniqueness) of a quotient $Q(x)$ and a remainder $R(x)$ such that

$$f(x) = Q(x)g(x) + R(x) \quad \text{and} \quad R(x) = 0 \text{ or } \deg(R) < \deg(g).$$

If the divisor is $g(x) = x - r$ then either $R(x) = 0$ or its degree is zero; in both cases, $R(x)$ is a constant that is independent of x; that is

$$f(x) = q(x)(x-r) + R.$$

Setting $x = r$ in this formula, we obtain:

$$f(r) = R$$

A slightly different proof, which may appear to some people as more elementary, starts with an observation that $f(x) - f(r)$ is a linear combination of terms of the form $x^k - r^k$ each of which is divisible by $x - r$ since $x^k - r^k = (x - r)(x^{k-1} + x^{k-2}r + \ldots + xr^{k-2} + r^{k-1})$.

Applications

The polynomial remainder theorem may be used to evaluate $f(r)$ by calculating the remainder, R. Although polynomial long division is more difficult than evaluating the function itself, synthetic division is computationally easier. Thus, the function may be more "cheaply" evaluated using synthetic division and the polynomial remainder theorem.

The factor theorem is another application of the remainder theorem: if the remainder is zero, then the linear divisor is a factor. Repeated application of the factor theorem may be used to factorize the polynomial.

FACTOR THEOREM

In algebra, the factor theorem is a theorem linking factors and zeros of a polynomial. It is a special case of the polynomial remainder theorem.

The factor theorem states that a polynomial $f(x)$ has a factor $(x - k)$ if and only if $f(k) = 0$ (i.e. k is a root).

Factorization of Polynomials

Two problems where the factor theorem is commonly applied are those of factoring a polynomial and finding the roots of a polynomial equation; it is a direct consequence of the theorem that these problems are essentially equivalent.

The factor theorem is also used to remove known zeros from a polynomial while leaving all unknown zeros intact, thus producing a lower degree polynomial whose zeros may be easier to find. Abstractly, the method is as follows:

1. "Guess" a zero a of the polynomial f. (In general, this can be *very hard*, but maths textbook problems that involve solving a polynomial equation are often designed so that some roots are easy to discover.)

2. Use the factor theorem to conclude that $(x - a)$ is a factor of $f(x)$.

3. Compute the polynomial $g(x) = f(x) / (x - a)$, for example using polynomial long division or synthetic division.

4. Conclude that any root $x \neq a$ of $f(x) = 0$ is a root of $g(x) = 0$. Since the polynomial degree of g is one less than that of f, it is "simpler" to find the remaining zeros by studying g.

Example:

Find the factors of

$$x^3 + 7x^2 + 8x + 2$$

To do this one would use trial and error (or the rational root theorem) to find the first x value that causes the expression to equal zero. To find out if $(x-1)$ is a factor, substitute $x = 1$ into the polynomial above:

$$x^3 + 7x^2 + 8x + 2 = (1)^3 + 7(1)^2 + 8(1) + 2$$

$$= 1 + 7 + 8 + 2$$

$$= 18.$$

As this is equal to 18 and not 0 this means $(x-1)$ is not a factor of $x^3 + 7x^2 + 8x + 2$. So, we next try $(x+1)$ (substituting $x = -1$ into the polynomial):

$$(-1)^3 + 7(-1)^2 + 8(-1) + 2.$$

This is equal to 0. Therefore $x-(-1)$, which is to say $x+1$, is a factor, and -1 is a root of $x^3 + 7x^2 + 8x + 2$.

The next two roots can be found by algebraically dividing $x^3 + 7x^2 + 8x + 2$ by $(x+1)$ to get a quadratic:

$$\frac{x^3 + 7x^2 + 8x + 2}{x+1} = x^2 + 6x + 2,$$

and therefore $(x+1)$ and $x^2 + 6x + 2$ are factors of $x^3 + 7x^2 + 8x + 2$. Of these the quadratic factor can be further factored using the quadratic formula, which gives as roots of the quadratic $-3 \pm \sqrt{7}$. Thus the three irreducible factors of the original polynomial are $x+1$, $x-(-3+\sqrt{7})$ and $x-(-3-\sqrt{7})$.

ABEL–RUFFINI THEOREM

In algebra, the Abel–Ruffini theorem (also known as Abel's impossibility theorem) states that there is no general algebraic solution—that is, solution in radicals— to polynomial equations of degree five or higher. The theorem is named after Paolo Ruffini, who made an incomplete proof in 1799, and Niels Henrik Abel, who provided a proof in 1823. Évariste Galois independently proved the theorem in a work that was posthumously published in 1846.

Interpretation

The content of this theorem is frequently misunderstood. It does not assert that higher-degree polynomial equations are unsolvable. In fact, the opposite is true: every non-constant polynomial equation in one unknown, with real or complex coefficients, has at least one complex number as solution; this is the fundamental theorem of algebra. Although the solutions cannot always be expressed exactly with radicals, they can be computed to any desired degree of accuracy using numerical methods such as the Newton–Raphson method or Laguerre method, and in this way they are no different from solutions to polynomial equations of the second, third, or fourth degrees.

The theorem only concerns the form that such a solution must take. The theorem says that not all solutions of higher-degree equations can be obtained by starting with the equation's coefficients and rational constants, and repeatedly forming sums, differences, products, quotients, and radicals (n-th roots, for some integer n) of previously obtained numbers. This clearly excludes the possibility of having any formula that expresses the solutions of an arbitrary equation of degree 5 or higher in terms of its coefficients, using only those operations, or even of having different formulas for different roots or for different classes of polynomials, in such a way as to cover all cases. (In principle one could imagine formulas using irrational numbers as constants, but even if a finite number of those were admitted at the start, not all roots of higher-degree equations could be obtained.) However some polynomial equations, of arbitrarily high degree, are solvable with such operations. Indeed, if the roots happen to be rational numbers, they can trivially be expressed as constants. The simplest nontrivial example is the equation $x^n = a$, where a is a positive real number, which has n solutions, given by:

$$x = \sqrt[n]{a}.e^{i2\pi k/n}, k = 0,1,...,n-1.$$

Here the expression $e^{i2\pi k/n}$ which appears to involve the use of the exponential function, in fact just gives the different possible values of $\sqrt[n]{1}$ (the n-th roots of unity), so it involves only extraction of radicals.

Lower-degree Polynomials

The solutions of any second-degree polynomial equation can be expressed in terms of addition, subtraction, multiplication, division, and square roots, using the familiar quadratic formula: The roots of the following equation are shown below:

$$ax^2 + bx + c = 0, \ a \neq 0$$

$$x = \frac{-b \pm \sqrt{b^2 - 4ac}}{2a}.$$

Analogous formulas for third- and fourth-degree equations, using cube roots and fourth roots, had been known since the 16th century.

Quintics and Higher

The Abel–Ruffini theorem says that there are some fifth-degree equations whose solution cannot be so expressed. The equation $x^5 - x + 1 = 0$ is an example. Some other fifth degree equations can be solved by radicals, for example $x^5 - x^4 - x + 1 = 0$ which factorizes to $(x-1)(x-1)(x+1)(x+i)(x-i) = 0$. The precise criterion that distinguishes between those equations that can be solved by radicals and those that cannot was given by Évariste Galois and is now part of Galois theory: a polynomial equation can be solved by radicals if and only if its Galois group (over the rational numbers, or more generally over the base field of admitted constants) is a solvable group.

Today, in the modern algebraic context, we say that second, third and fourth degree polynomial equations can always be solved by radicals because the symmetric groups S_2, S_3 and S_4 are solvable groups, whereas S_n is not solvable for $n \geq 5$. This is so because for a polynomial of degree n with indeterminate coefficients (i.e., given by symbolic parameters), the Galois group is the full symmetric group S_n (this is what is called the "general equation of the n-th degree"). This remains true if the coefficients are concrete but algebraically independent values over the base field.

Proof:

The following proof is based on Galois theory. Historically, Ruffini and Abel's proofs precede Galois theory.

One of the fundamental theorems of Galois theory states that an equation is solvable in radicals if and only if it has a solvable Galois group, so the proof of the Abel–Ruffini theorem comes down to computing the Galois group of the general polynomial of the fifth degree.

Let y_1 be a real number transcendental over the field of rational numbers Q and let y_2 be a real number transcendental over $Q(y_1)$ and so on to y_5 which is transcendental over $Q(y_1, y_2, y_3, y_4)$ These numbers are called independent transcendental elements over Q. Let $E = Q(y_1, y_2, y_3, y_4, y_5)$ and let

$$f(x) = (x - y_1)(x - y_2)(x - y_3)(x - y_4)(x - y_5) \in E = [x].$$

Multiplying $f(x)$ out yields the elementary symmetric functions of the y_n:

$$s_1 = y_1 + y_2 + y_3 + y_4 + y_5$$
$$s_2 = y_1 y_2 + y_1 y_3 + y_1 y_4 + y_1 y_5 + y_2 y_3 + y_2 y_5 + y_3 y_4 + y_3 y_5 + y_4 y_5$$
$$s_3 = y_1 y_2 y_3 + y_1 y_2 y_4 + y_1 y_2 y_5 + y_1 y_3 y_4 + y_1 y_3 y_5 + y_1 y_4 y_5 + y_2 y_3 y_4 + y_2 y_3 y_5 + y_2 y_4 y_5 + y_3 y_4 y_5$$
$$s_4 = y_1 y_2 y_3 y_4 + y_1 y_2 y_3 y_5 + y_1 y_2 y_4 y_5 + y_1 y_3 y_4 y_5 + y_2 y_3 y_4 y_5$$
$$s_5 = y_1 y_2 y_3 y_4 y_5.$$

The coefficient of x^n in $f(x)$ is thus $(-1)^{5-n} s_{5-n}$. Because our independent transcendentals y_n act as indeterminates over Q, every permutation σ in the symmetric group on 5 letters S_5 induces an automorphism σ' on E that leaves Q fixed and permutes the elements y_n. Since an arbitrary rearrangement of the roots of the product form still produces the same polynomial, e.g.:

$$(y - y_3)(y - y_1)(y - y_2)(y - y_5)(y - y_4)$$

is still the same polynomial as

$$(y - y_1)(y - y_2)(y - y_3)(y - y_4)(y - y_5)$$

the automorphisms σ' also leave E fixed, so they are elements of the Galois group $G(E / Q)$ Now, since $|S_5| = 5!$ it must be that $|G(E / Q)| \geq 5!$ as there could possibly be automorphisms there that are not in S_5 However, since the relative automorphisms Q for splitting field of a quintic polynomial has at most 5! elements, $|G(E/Q)| = 5!$ and so $G(E/Q)$ must be isomorphic to S_5 Generalizing this argument shows that the Galois group of every general polynomial of degree n is isomorphic to S_n

And what of S_5? The only composition series of S_5 is $S_5 \geq A_5 \geq \{e\}$ (where A_5 is the alternating group on five letters, also known as the icosahedral group). However, the quotient group $A_5 / \{e\}$ (isomorphic to A_5 itself) is not an abelian group, and so S_5 is not solvable, so it must be that the general polynomial of the fifth degree has no solution in radicals. Since the first nontrivial normal subgroup of the symmetric group on n letters is always the alternating group on n letters, and since the alternating groups on n letters for $n \geq 5$ are always simple and non-abelian, and hence not solvable, it also says that the general polynomials of all degrees higher than the fifth also have no solution in radicals.

Note that the above construction of the Galois group for a fifth degree polynomial only applies to the general polynomial, specific polynomials of the fifth degree may have different Galois groups with quite different properties, e.g. $x^5 - 1$ has a splitting field generated by a primitive 5th root of unity, and hence its Galois group is abelian and the equation itself solvable by radicals. However, since the result is on the general polynomial, it does say that a general "quintic formula" for the roots of a quintic using only a finite combination of the arithmetic operations and radicals in terms of the coefficients is impossible.

AMITSUR–LEVITZKI THEOREM

In algebra, the Amitsur–Levitzki theorem states that the algebra of n by n matrices satisfies a certain identity of degree $2n$. It was proved by Amitsur and Levitsky. In particular matrix rings are polynomial identity rings such that the smallest identity they satisfy has degree exactly $2n$.

The standard polynomial of degree n is

$$S_n(x_1,\ldots,x_n) = \sum_{\sigma \in S_n} \text{sgn}(\sigma) x_{\sigma(1)} \cdots x_{\sigma(n)}$$

in non-commutative variables $x_1,\ldots,x_n$, where the sum is taken over all $n!$ elements of the symmetric group S_n.

The Amitsur–Levitzki theorem states that for n by n matrices $A_1,\ldots,A_{2n}$ then

$$S_{2n}(A_1,\ldots,A_{2n}) = 0.$$

Proofs:

Amitsur and Levitzki gave the first proof.

Kostant deduced the Amitsur–Levitzki theorem from the Koszul–Samelson theorem about primitive cohomology of Lie algebras.

Swan and Swan gave a simple combinatorial proof as follows. By linearity it is enough to prove the theorem when each matrix has only one nonzero entry, which is 1. In this case each matrix can be encoded as a directed edge of a graph with n vertices. So all matrices together give a graph on n vertices with $2n$ directed edges. The identity holds provided that for any two vertices A and B of the graph, the number of odd Eulerian paths from A to B is the same as the number of even ones. (Here a path is called odd or even depending on whether its edges taken in order give an odd or even permutation of the $2n$ edges.) Swan showed that this was the case provided the number of edges in the graph is at least $2n$, thus proving the Amitsur–Levitzki theorem.

Razmyslov gave a proof related to the Cayley–Hamilton theorem.

Rosset gave a short proof using the exterior algebra of a vector space of dimension $2n$.

Procesi gave another proof, showing that the Amitsur–Levitzki Theorem is the Cayley–Hamilton identity for the generic Grassman matrix.

BERNSTEIN–KUSHNIRENKO THEOREM

Bernstein–Kushnirenko theorem (also known as BKK theorem or Bernstein–Khovanskii–Kushnirenko theorem), proven by David Bernstein and Anatoli Kushnirenko [ru] in 1975, is a theorem in algebra. It states that the number of non-zero complex solutions of a system of Laurent polynomial equations $f_1 = \cdots = f_n = 0$ is equal to the mixed volume of the Newton polytopes of the polynomials $f_1, \ldots, f_n$, assuming that all non-zero coefficients of f_n are generic. A more precise statement is as follows:

Theorem Statement

Let A be a finite subset of $\mathbb{Z}^n$ Consider the subspace L_A of the Laurent polynomial algebra $\mathbb{C}\left[x_1^{\pm 1}, \ldots, x_n^{\pm 1}\right]$ consisting of Laurent polynomials whose exponents are in A. That is:

$$L_A = \left\{ f \mid f(x) = \sum_{\alpha \in A} c_\alpha x^\alpha, c_\alpha \in \mathbb{C} \right\},$$

where for each $\alpha = (a_1, \ldots, a_n) \in \mathbb{Z}^n$ we have used the shorthand notation x^α to denote the monomial $x_1^{a_1} \cdots x_n^{a_n}$.

Now take n finite subsets $A_1, \ldots, A_n$ with the corresponding subspaces of Laurent polynomials $L_{A_1}, \ldots, L_{A_n}$. Consider a generic system of equations from these subspaces, that is:

$$f_1(x) = \cdots = f_n(x) = 0,$$

where each f_i is a generic element in the (finite dimensional vector space) L_{A_i}.

The Bernstein–Kushnirenko theorem states that the number of solutions $x \in (\mathbb{C} \setminus 0)^n$ of such a system is equal to

$$n! V(\Delta_1, \ldots, \Delta_n),$$

where V denotes the Minkowski mixed volume and for each i, Δ_i is the convex hull of the finite set of points A_i. Clearly Δ_i is a convex lattice polytope. It can be interpreted as the Newton polytope of a generic element of the subspace L_{A_i}.

In particular, if all the sets A_i are the same $A = A_1 = \cdots = A_n$, then the number of solutions of a generic system of Laurent polynomials from L_A is equal to

$$n!\,\mathrm{vol}(\Delta),$$

where Δ is the convex hull of A and vol is the usual n-dimensional Euclidean volume. Note that even though the volume of a lattice polytope is not necessarily an integer, it becomes an integer after multiplying by $n!$.

CARTAN–BRAUER–HUA THEOREM

In abstract algebra, the Cartan–Brauer–Hua theorem (named after Richard Brauer, Élie Cartan, and Hua Luogeng) is a theorem pertaining to division rings. It says that given two division rings $K \subseteq D$ such that xKx^{-1} is contained in K for every x not equal to 0 in D, either K is contained in the center of D, or $K = D$. In other words, if the unit group of K is a normal subgroup of the unit group of D, then either $K = D$ or K is central

Proof:

Any Element Inside Commutes with any Element Outside

For nonzero elements $x, y \in L^*$, we denote by $[x, y]$ the *multiplicative* commutator $xyx^{-1}y^{-1}$ and by $c_x(y)$ the element xyx^{-1}.

We denote by c_x the map $y \mapsto xyx^{-1}$. Here, $x, \in L^*$ but y is allowed to be zero.

Given: $g \in K^*$ and $a \in L \setminus K$.

To prove: $[g, a] = 1$.

Proof: The key idea is to play off the additive and the multiplicative structure against each other, and use the fact that the map $y \mapsto xyx^{-1}$ is an automorphism of both the additive and the multiplicative structure.

Step no.	Assertion/construction	Given data/ assumptions used	Previous steps used	Explanation	Commentary
1	$g, a, a+1$ are all in L^*, so the notations c_g, $[g,a]$, and $[g, a+1]$ make sense.	$g \in K^*$, $a \in L \setminus K$.		$g \in K^*$, so $g \in L^*$. Further, $a \in L \setminus K$, so a is nonzero too. Further, since $a \notin K$, we know that $a \neq -1$ (since -1 would be in any skew field) and hence $a+1 \neq 0$.	The choice of a and $a+1$ allows us to play on addition.

2	$[g,a] \in K^*$ and $[g, a+1] \in K^*$.	K^* is normal in L^*	Step (1)	The commutator of any element in a normal subgroup with an element in the whole group lies in the normal subgroup. This is one of the equivalent definitions of normal subgroup.	
3	$c_g(a) = [g, a]a$ and $c_g(a+1) = [g, a+1](a+1)$		Step (1)	Just follows from definitions: $[g, a] = gag^{-1}a^{-1}$ and $c_g(a) = gag^{-1}$. The same logic applies replacing a by $a+1$.	The multiplicative commutator is not convenient because it is not additive/linear in either variable. So, we rewrite it in terms of c_g, which preserves the additive structure.
4	$c_g(a+1) = c_g(a) + 1$		Step (1)	Just follows from definitions: $c_g(a+1) = g(a+1)g^{-1} =$, $gag^{-1} + g1g^{-1} = gag^{-1} + 1$ which is $c_g(a) + 1$.	More manipulation.
5	$[g, a+1](a+1) = [g, a]a + 1$, or equivalently, $([g, a+1] - [g, a])a = 1 - [g, a+1]$		Steps (3) and (4)	We plug in expressions for $c_g(a+1)$ and $c_g(a)$ from step (3) into the expression for step (4).	More manipulation.
6	The assumption $[g, a+1] \neq [g, a]$ would lead to a contradiction, hence we must have $[g, a+1] = [g, a]$.	$a \notin K$. Also, every nonzero element is invertible because L is a skew field.	Steps (2), (5)	If $[g, a+1] \neq [g, a]$, then the difference $[g, a+1] - [g, a]$ is invertible. Multiplying the second formulation of step (5) on the left by the inverse of $[g, a+1] - [g, a]$ and simplifying, we get $a = ([g, a+1] - [g, a])^{-1}(1 - [g, a+1])$. Both $[g, a+1]$ and $[g, a]$ are in K^* by step (2). Thus, the right side is an expression involving terms in K and hence must be in K. This contradicts the assumption that $a \notin K$.	More manipulation.
7	Plugging $[g, a+1] = [g, a]$ in the result of step (5) gives $[g, a] = 1$.		Step (5)	We get $[g, a](a+1) = [g, a]a + 1$. Cancel out $[g, a]a$ additively form both sides, and we get $[g, a] = 1$.	

The Finishing Touch

Now, if K is a *proper* subset of L, we will show that K^* is contained inside the center. We already know that every element of K^* commutes with every element of $L \setminus K$. So it suffices to show that any two elements of K^* commute.

Let $g, h \in K^*$. Then take any $a \in L \setminus K$. Then, $a + h \in L \setminus K$. Thus, g commutes with both $a + h$ and a. Hence g must commute with the difference, which is h.

HILBERT'S BASIS THEOREM

Hilbert's Basis Theorem is a result concerning Noetherian rings. It states that if A is a (not necessarily commutative) Noetherian ring, then the ring of polynomials $A[x_1, x_2, \ldots, x_n]$ is also a Noetherian ring. (The converse is evidently true as well).

Note that n must be finite; if we adjoin infinitely many variables, then the ideal generated by these variables is not finitely generated.

The theorem is named for David Hilbert, one of the great mathematicians of the late nineteenth and twentieth centuries. He first stated and proved the theorem in 1888, using a nonconstructive proof that led Paul Gordan to declare famously, "Das ist nicht Mathematik. Das ist Theologie. [This is not mathematics. This is theology.]" In time, though, the value of nonconstructive proofs was more widely recognized.

Proof

By induction, it suffices to show that if A is a Noetherian ring, then so is $A[x]$.

To this end, suppose that $\mathfrak{a}_0 \subset \mathfrak{a}_1 \subset \cdots$ is an ascending chain of (two-sided) ideals of $A[x]$.

Let $\mathfrak{c}_{i,j}$ denote the set of elements a of A such that there is a polynomial in $\mathfrak{a}_i$ with degree at most i and with a as the coefficient of x^j. Then $\mathfrak{c}_{i,j}$ is a two-sided ideal of A; furthermore, for any $i' \geq i$, $j' \geq j$,

$$\mathfrak{c}_{i,j} \subset \mathfrak{c}_{i',j}, \mathfrak{c}_{i,j'}$$

Since A is Noetherian, it follows that for every ≥ 0, the chain

$$\mathfrak{c}_{i,0} \subset \mathfrak{c}_{i,1} \subset \cdots$$

Stabilizes to some ideal $\mathfrak{m}_i$. Furthermore, the ascending chain

$$\mathfrak{m}_0, \mathfrak{m}_1, \ldots$$

also stabilizes to some ideal $\mathfrak{m} = \mathfrak{c}_{A,B}$. Then for any $i \geq A$ and any $j \geq 0$,

$$\mathfrak{c}_{i,j} = \mathfrak{c}_{A,j}.$$

We claim that the chain $(\mathfrak{a}_k)_{k=0}^{\infty}$ stabilizes at $\mathfrak{a}_A$. For this, it suffices to show that for all $k \geq A$, $\mathfrak{a}_k \subset \mathfrak{a}_A$. We will thus prove that all polynomials of degree n in $\mathfrak{a}_k$ are also elements of $\mathfrak{a}_A$, by induction on n.

For our base case, we note that $\mathfrak{c}_{k,0} = \mathfrak{c}_{M,0}$, and these ideals are the sets of degree-zero polynomials in $\mathfrak{a}_k$ and $\mathfrak{a}_M$, respectively.

Now, suppose that all of $\mathfrak{a}_k$'s elements of degree n - 1 or lower are also elements of $\mathfrak{a}_M$. Let p be an element of degree n in $\mathfrak{a}_k$. Since

$$\mathfrak{c}_{k,n} = \mathfrak{c}_{A,n},$$

there exists some element $q \in \mathfrak{a}_A$ with the same leading coefficient as p. Then by inductive hypothesis,

$$p - q \in \mathfrak{a}_A,$$

So

$$p \in \mathfrak{a}_A,$$

as desired.

References

- Fundamental-theorem-of-algebra: brilliant.org, Retrieved 06 August, 2019

- Cox, David A.; Little, John; O'Shea, Donal (2005). Using algebraic geometry. Graduate Texts in Mathematics. 185 (Second ed.). Springer. ISBN 0-387-20706-6. MR 2122859

- Arnold, Vladimir; et al. (2007). "Askold Georgievich Khovanskii". Moscow Mathematical Journal. 7 (2): 169–171. MR 2337876

- Cartan-Brauer-Hua-theorem: groupprops.subwiki.org, Retrieved 22 May, 2019

- Lam, Tsit-Yuen (2001). A First Course in Noncommutative Rings (2nd ed.). Berlin, New York: Springer-Verlag. ISBN 978-0-387-95325-0. MR 1838439

We would like to thank the editorial team for lending their expertise to make the book truly unique. They have played a crucial role in the development of this book. Without their invaluable contributions this book wouldn't have been possible. They have made vital efforts to compile up to date information on the varied aspects of this subject to make this book a valuable addition to the collection of many professionals and students.

This book was conceptualized with the vision of imparting up-to-date and integrated information in this field. To ensure the same, a matchless editorial board was set up. Every individual on the board went through rigorous rounds of assessment to prove their worth. After which they invested a large part of their time researching and compiling the most relevant data for our readers.

The editorial board has been involved in producing this book since its inception. They have spent rigorous hours researching and exploring the diverse topics which have resulted in the successful publishing of this book. They have passed on their knowledge of decades through this book. To expedite this challenging task, the publisher supported the team at every step. A small team of assistant editors was also appointed to further simplify the editing procedure and attain best results for the readers.

Apart from the editorial board, the designing team has also invested a significant amount of their time in understanding the subject and creating the most relevant covers. They scrutinized every image to scout for the most suitable representation of the subject and create an appropriate cover for the book.

The publishing team has been an ardent support to the editorial, designing and production team. Their endless efforts to recruit the best for this project, has resulted in the accomplishment of this book. They are a veteran in the field of academics and their pool of knowledge is as vast as their experience in printing. Their expertise and guidance has proved useful at every step. Their uncompromising quality standards have made this book an exceptional effort. Their encouragement from time to time has been an inspiration for everyone.

The publisher and the editorial board hope that this book will prove to be a valuable piece of knowledge for students, practitioners and scholars across the globe.

INDEX

www.ingramcontent.com/pod-product-compliance
Lightning Source LLC
Chambersburg PA
CBHW082012190326
41458CB00010B/3159